Hydro Power Developments

Fluid Machinery Committee

Seminar Advisers

IMechE
Seminar Publication

I MECH E

150th Anniversary

1847 - 1997

Hydro Power Developments
– Current Projects, Rehabilitation, and Power Recovery

30 October 1997

Organized by the
Fluid Machinery Committee of
the Power Industries Division of
The Institution of Mechanical Engineers

IMechE Seminar Publication 1997–4

Published by Mechanical Engineering Publications Limited for
The Institution of Mechanical Engineers, Bury St Edmunds and London, UK.

First Published 1997

ISSN 1357–9193
ISBN 1 86058 121 8

A CIP catalogue record for this book is available from the British Library.

Printed by The Ipswich Book Company, Suffolk, UK.

Contents

Related Titles of Interest

Title	Editor	ISBN
Energy Saving in the Design and Operation of Pumps	IMechE Seminar 1996–1	1 86058 030 0
Water Pipeline Systems	R Chilton (BHR Group Conference Publication No. 23)	1 86058 088 2
Pressure Surges and Fluid Transients in Pipelines and Open Channels	A Boldy (BHR Group Conference Publication No. 19)	0 85298 991 1

For the full range of titles published by MEP contact:

Sales Department
Mechanical Engineering Publications Limited
Northgate Avenue
Bury St Edmunds
Suffolk
IP32 6BW
UK

Tel: 01284 763277
Fax: 01284 704006

Llyn Brianne hydro-electric scheme

R C A TAYLOR BSc, CEng, MIMechE, MCIWEM
Hyder Industrial Limited, Cardiff, UK

SYNOPSIS

This paper gives an account of the development of the hydroelectric generating station at Llyn Brianne, together with the associated raising of Brianne dam.

Operational constraints are considered particularly with respect to water resource requirements and environmental issues.

The complete installation is described in detail, as it was constructed including all plant, equipment and control systems.

1. BACKGROUND

Llyn Brianne reservoir is situated in South West Wales about twenty kilometres north of Llandovery. The dam was built by Welsh Water in the early seventies and was intended solely for the purposes of river regulation to support drinking water abstraction further downstream. Dwr Cymru Welsh Water is the main Water Supply Utility Company in Wales and now forms part of the Hyder plc Group.

The hydropower potential of the site was recognised right from the start, but this potential was never realised because of the distance and high cost of the connection to the local electricity distribution network.

Things changed when in 1990 the Government announced the Non-Fossil Fuel Obligation (NFFO) Renewables Orders. This more or less coincided with the privatisation of the water industry which in turn stimulated a new impetus to none-core activities generally, and in particular to fully exploit existing assets. Hydropower was seen by Welsh Water as a diversification which was closely allied to the main business both technically and commercially. It is felt that the engineering, construction and operational skills necessary for hydro-electric power development have a very close similarity to the pipework and pumping systems inherent in a large water supply undertaking. This position was further strengthened by the acquisition by Welsh Water of Swalec - the main Welsh Regional Electricity Company. It is also recognised that the income stream from hydro is continuous and secure well into the future which again compares well with the water and electricity businesses.

There are many dams throughout Wales and indeed throughout the whole of the UK which are of the order of 30 metres high. Brianne stands out straight away as it is 85 metres high. Hence for a given catchment and rainfall the power output from Brianne is almost three times usual potential. Brianne catchment is 88 km^2 and average annual rainfall is 1900 mm.

In January 1995 Welsh Water Industrial Services (now Hyder Industrial) were successful in winning a 4.35 MW power supply contract under the third tranche of the NFFO. The contract was won on a competitive basis and is a fixed price. The NFFO is a mechanism whereby a premium rate is paid for the power generated. Potential developers submit competitive bids for particular projects and each bid is assessed in comparison with others in the same technology band i.e. hydro, wind, landfill gas etc. The bid value represents the price per kilowatt hour that would give a viable return on the initial development costs. Contracts have been awarded to the lowest bid prices in each band up to a level where the aggregate of awarded contracts reaches the target generation capacity. Power generation contracts are for fifteen years and there is no capital incentive - all development costs have to be funded from the eventual generation income.

2. GRID CONNECTION ISSUES

The dam is situated 20 km from the nearest connection point to the local electricity distribution network. and the original contract bid price had been based on a grid connection charge quoted by the local Regional Electricity Company (Swalec). This quotation incorporated an overhead transmission line. Although planning permission was initially granted by the local Planning Authority for overhead lines, this permission was subsequently withdrawn following fierce opposition from residents of the upper Towy Valley who would not tolerate the overhead wire and wooden pole line.

The landscape of the Towy Valley is undoubtedly of very high quality and generally becomes more remote and spectacular from South to North as the valley passes into the Cambrian Mountains. The valley encompasses designated Special Landscape Areas and Areas of Great Landscape Value together with Sites of Special Scientific Interest and Environmentally Sensitive Areas.

For a time it seemed like the project might not go ahead because of the significant additional cost of laying 20 km of underground cable. The only alternative appeared to be a drawn out and costly Public Enquiry.

However, a solution was found which enabled the extra capital to be supported by increased generation, - through raising the top water level of the dam and seeking to reduce the costs of undergrounding the cable.

3. DAM RAISING

There are three categories of discharge from the reservoir from which we can generate power:

3.1 **Compensation**
A quantity of about 0.8 cumecs is released at all times - 24 hours /day, 365 days /year to provide a basic minimum flow for the river.

3.2 **Augmentation**
A quantity of about 2.8 cumecs is released during dry periods to match the abstractions for water supply further down the river.

3.3 **Overspill**
During the winter when the reservoir is full, excess water overflows and runs down the spillway. During wet weather this quantity of overspill is typically greater than 15 cumecs.

The quantity of water which can pass through the turbines is limited by the size of the machines (maximum throughout approximately 8 cumecs with all turbines running). The solution involved raising the top water level of the reservoir such that we are able to 'catch' a large proportion of the water which would otherwise spill and be lost to hydro generation - thus lowering the level again, ready to 'catch' the next rainstorm event.

The dam was originally designed to be raised by a maximum of 18 metres. By coincidence, in 1991 Welsh Water commissioned a study to assess the feasibility of raising both Llyn Brianne and Llys y Fran dams. Llys y Fran is a much smaller dam situated in West Wales, and was raised as a result of this study. However, Brianne raising did not go ahead as there was no immediate need for the additional water resource. The study for Brianne examined three options ie raising the dam by one, two or the full eighteen metres.

We carried out computer modelling to investigate the effect on generation of each of these three scenarios. It was found that the increased generation from raising the dam by just one metre was sufficient not only to cover the additional cost of undergrounding the cable, but also to cover the cost of raising the dam. The returns for raising the dam higher were obviously increased but not by a sufficient amount to service the additional capital required, so the one metre figure was taken as an optimum value.

Figures 1 and 2 show the modelled generation profiles for 1980 - which is taken as a typical year. Scenario B1 (figure 1) represents the scheme as originally designed and Scenario I (figure 2) represents the effect of raising the dam by one metre and utilising a two metre drawdown from the new top water level.

The shaded areas represent the quantity and duration of spill. Line 2 represents releases from the reservoir which can be used for generation. Lines 3 and 4 represent the Lower and Upper reservoir rule curves and line 5 represents the actual storage volume throughout this year.

A great deal of work went into assessing the generation returns and environmental impacts of the various possible operating scenarios. Variables considered not only included differing dam heights but also possible changes to abstraction releases and necessary mitigating measures.

Having reached a satisfactory solution technically it was then necessary to sell the idea not only to the local residents but also to a significant number of consultees and interested parties. The reservoir covered three separate planning areas, a new Impoundment License was required together with land drainage consent from the Environment Agency and an existing road bridge was also affected.

The mechanism for raising the dam was quite straightforward. The top water level is dictated by the height of the spillway - and this was simply raised in concrete by the requisite amount. The spillway being 50 m wide. (figure 3)

The dam itself is of clay core, rock fill construction, and although it was already sufficiently high for the proposed new top water level, the clay core did not extend to the top of the dam and had to be modified. (figure 4)

The crest of the dam is approximately 290 m long and carries a roadway. It was necessary to excavate a trench along the whole length of the dam 2.5 m deep down into the clay core. A one metre wide concrete 'key' was then cast up to the original road level and following back filling and compaction the road surface was replaced. A certain amount of additional rip rap was also required but the overall aspect of the dam was largely unchanged.

All the land surrounding the reservoir up to an elevation of 3 m above the old top water level was owned by Welsh Water, so the additional inundation did not affect other landowners. A number of people did have right of access across the dam however and during the raising works (approx. 4 months) the road had to be closed. The needs of each user was assessed and measures put in place as appropriate.

4. OPERATIONAL CONSTRAINTS

Much of the liaison work involved environmental issues and the Environment Agency laid down very strict operating rules to safeguard the various areas of concern. Separate rules

© IMechE 1997 S487/00

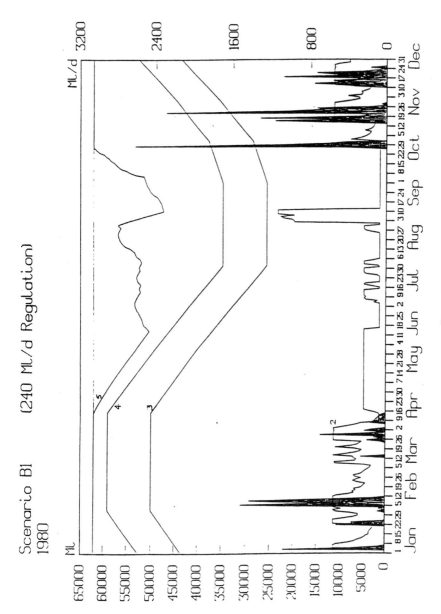

Scenario B1   (240 ML/d Regulation)
1980

Figure 1

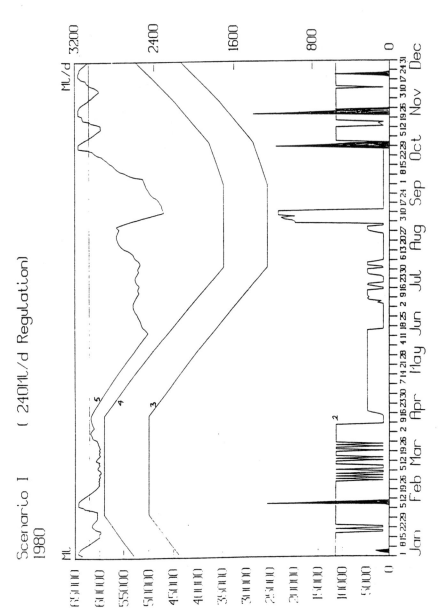

Scenario 1 (2401ML/d Regulation)
1980

Figure 2

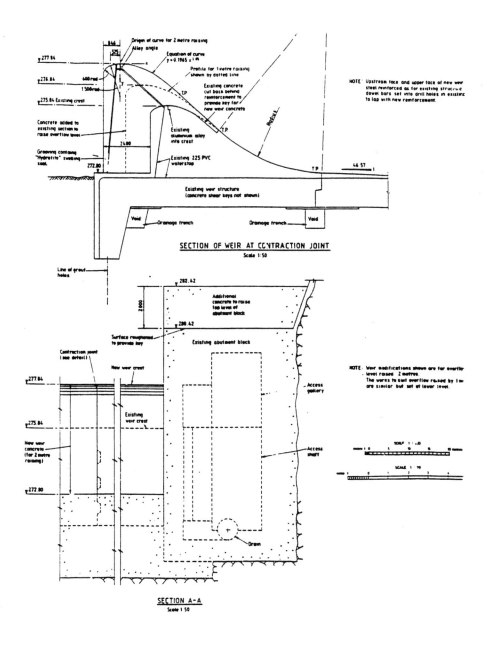

SECTION OF WEIR AT CONTRACTION JOINT
Scale 1:50

SECTION A-A
Scale 1:50

Figure 3

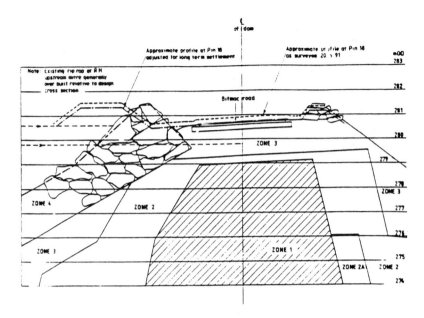

CROSS SECTION OF CREST
AT R.H.END OF DAM (NEAR PIN 18)

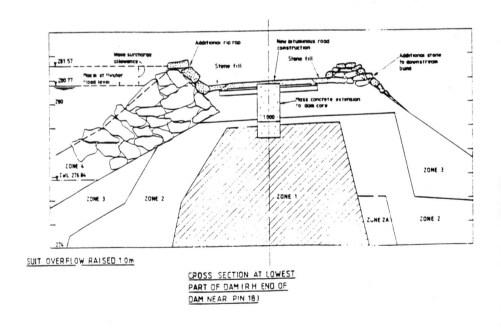

SUIT OVERFLOW RAISED 1.0m

CROSS SECTION AT LOWEST
PART OF DAM (R H END OF
DAM NEAR PIN 18)

Figure 4

exist for the summer and winter periods. During the summer (6 April to 1 November) releases are restricted to compensation and augmentation only. Augmentation is controlled by a river flow level at Nantgaredig which is approx. 40 km below Brianne. In short, augmentation is only necessary after about a week of dry weather.

During the environmental appraisal it was found that the existing disperser valves had the effect of raising the river temperature during warm weather and also lowering the river temperature during frosty weather.

It is known that young salmon (salmonids) do not feed below 7 °C and during warm spring days this critical temperature is only achieved by using the disperser. The reservoir (bottom) draw off temperature is below this level and direct release through the turbines can have the effect of preventing the fish from feeding.

Hence an operating rule was established as follows: During the months of May, June and July if the instantaneous river temperature is between 6 °C and 7 °C and the instantaneous air temperature is above 11.5 °C the disperser is used. In practice the available flow is shared between turbine and disperser - the proportion being varied until the target 7 °C is reached.

A key concern of the Environment Agency was the issue of acid rain. There has been a liming programme in operation at Llyn Brianne for a number of years, and this has been of major benefit to the fish population of the Towy. It has been realised that the high pH water from Brianne has the effect of diluting/neutralising the acid waters entering the Towy from tributaries below the dam, and in particular the Doethie. During the autumn when the problem is at its worst we have agreed to make specific hydro releases to match storm flows in the Doethie. In addition we have agreed to maintain and operate two newly installed lime dosing plants on the feeder rivers into Brianne.

After 1 November the generation is controlled by reservoir level. The rate of change of flow (and power output) mimics the flow changes caused by a natural rainstorm event, but is also designed to reduce overspill to a minimum.

The additional one metre of impoundment equates to three days generation at full capacity, but there will still be short periods when the reservoir overflows.

Obviously the main purpose of the dam is to support abstraction for drinking water and at all times the water supply requirements have to take precedence over hydro. Furthermore, it is imperative that the strategic storage capacity is not compromised and the reservoir must be full at the end of each winter to provide supplies through any drought period. With this in mind it has been agreed with both Welsh Water and the Environment Agency that a further one metre depth of storage may be used for hydro purposes during the winter. Hence from 1 November the target operating level of the reservoir is 2 m down from new top water level.

In order to guarantee as far as possible that the reservoir will be at the old top water level at the prescribed date of 15 March, the target operating level 'ramps-up' from 15 February to 15 March and flows are reduced to compensation only if necessary to achieve the required level.

5. CONSTRUCTION

Construction work on the site began in February 1996 and was hampered from the start not only by the inclement weather, but also the constant presence of water spray from the existing disperser valve releases.

The turbine house is situated at the foot of the dam adjacent to the existing valve house and involved the excavation of 3000 cubic metres of material. When the dam was first built, the rockfill material was obtained from the quarry adjacent. A certain amount of rip rap boulder stone was required for the dam raising exercise and it was logical to obtain this from the now abandoned quarry. Re-opening the quarry required a new planning consent, and a condition of this consent was that the quarry area must be landscaped. These landscaping works therefore provided an on site requirement for the turbine house excavated material and haulage off site was not necessary. (figure 5)

The turbine house is of traditional construction ie portal frame structure, concrete block walls and slate roof. The planning conditions required a pitched roof 'cottage effect' structure. The turbine hall is suspended over discharge chambers beneath floor level and a separate discharge chamber and weir is provided for each of the three turbines.

6.   MECHANICAL AND ELECTRICAL INSTALLATION

The original dam discharge pipework comprises a 66 inch steel outlet main which then subdivides into three branches, two at 54 inch and one at 36 inch. Each branch was equipped with a 30 inch access flange (blanked-off) and the turbine pipework is now fed from these three flanged connections.

In addition to the primary 66 inch discharge pipework, there is also a low level draw off scour pipe (36 inch). Use of the scour enabled flow to be maintained to the river during times when the 66 inch pipe was closed down for the interconnection works.

In this way, the existing strategic dam drawdown facilities were left intact. The original dam pipework was installed to civil engineering tolerances and the three connection flanges were out of alignment by 40 mm. The existing dimensions were measured simply but accurately (using piano wire) and a new branched manifold was workshop fabricated and craned into place - it fitted perfectly. This single fabrication converted three 30 inch connections to a 1350 mm pipe and weighs 12.5 tons. The decision to use all three access flange connections was strictly a commercial one - the pipe friction losses being directly proportional to reduced generation. (figure 6)

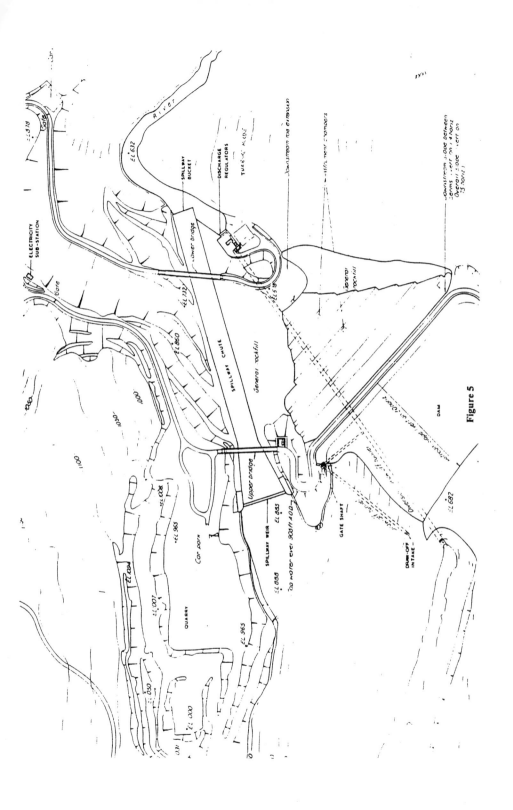

Figure 5

Manifold Elevation

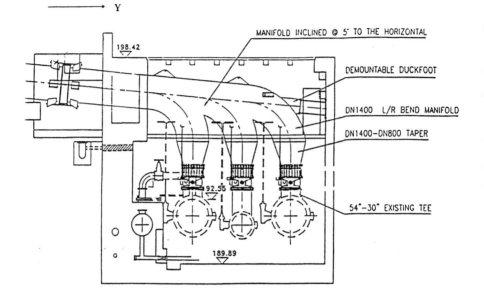

View in Y Direction

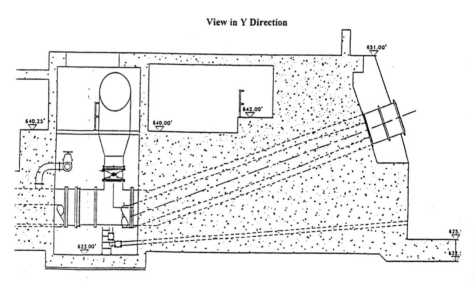

Figure 6

The plant comprises three 'Francis' turbines, two at 2.5 MW capacity and the third at 600 kw capacity. The 2.5 MW machines are larger than necessary, and this is due to utilising the manufacturers standard frame sizes. The machines are horizontally mounted with discharge vertically downward through the floor. The discharge chambers incorporate a weir which provides a uniform back pressure on the draft tubes. (figure 7)

The turbines have a very broad operating range, such that increasing the maximum inlet pressure by one metre did not necessitate any design changes.

The actual design of the turbines was left very much in the hands of the turbine manufacturers in order to ensure the most commercially competitive arrangement. The design chosen was of the 'overhung' runner configuration whereby the turbine runner is mounted directly onto the generator shaft and all thrust is accommodated by the generator bearings.

On the larger machines a bearing cooling system is incorporated. The bearings are oil lubricated and the generator shaft carries two oil pumps. The first simply pumps oil around the circuit from the oil reservoir to the bearing housing, and the second provides oil under pressure to drive a hydraulic motor driven fan and radiator cooling system.

The turbine volute casings are of cast iron and both turbine and generator are bolted down to a concrete plinth - There being no bedplate as such.

Maintenance facilities are good. There is ample room for machinery dismantling and the building incorporates at 10 ton overhead travelling crane. A hand operated crane was chosen for preference as infrequent use was anticipated.

The generation voltage is 660 volts - again this was a commercial decision relating to cost of generator manufacture. Power is then transformed on site to 33,000 volts for transmission via the 20 km long underground cable to the grid connection point in Llandovery.

Starting and stopping of turbines is achieved by hydraulic actuators and fail safe closedown is provided by compressed air reservoirs incorporated into the high pressure hydraulic circuits. Valve closing times are strictly managed to limit pressure surge in the event of a loss of load.

The main inlet valves are of the butterfly type and in addition to the hydraulic actuation a spring closure device is fitted.

Control is by programmable logic controller, and allows for local automatic or manual control together with remote manual override via a telemetry system to Welsh Water's 24 hour control centre.

The need to maintain a minimum flow to the river at all times of 0.8 cumecs is a statutory requirement, and so the control software has had to be configured to automatically open a disperser valve to a pre-set level in the event of all turbine flow ceasing for any reason.

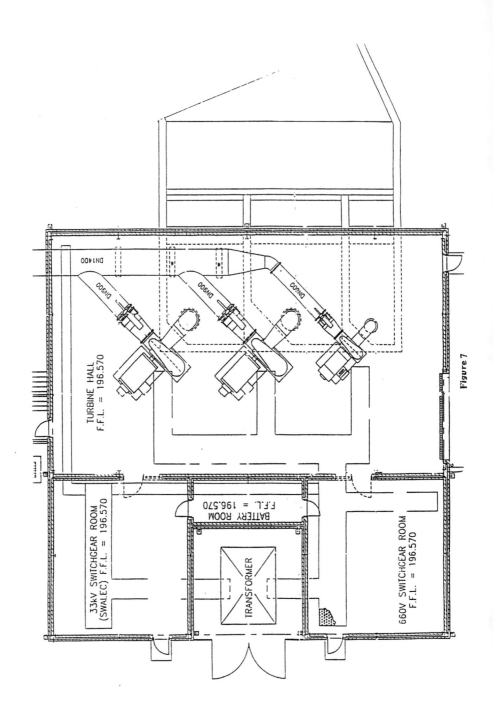

Figure 7

TURBINE HALL
F.F.L. = 196.570

33kV SWITCHGEAR ROOM
(SWALEC) F.F.L. = 196.570

BATTERY ROOM
F.F.L. = 196.570

TRANSFORMER

660V SWITCHGEAR ROOM
F.F.L. = 196.570

DN1400

DN900

DN900

DN400

The control equipment is supported by an Uninterruptible Power Supply and this has had to be of sufficient capacity to drive the disperser actuators.

Grid connection costs were driven down by competitive tendering independently of the Regional Electricity Company - although this did mean that such things as Streetworks Licences and Wayleave Agreements had to be negotiated by the developer rather than the electricity company. Much of the cable route was either in or adjacent to the public highway and at the top of the valley involved excavation in rock. Lower down the valley excavation was easier and it was possible to lay the cable on the field side of the hedge. A route was found around Llandovery thus avoiding major disruption to traffic.

7. PERFORMANCE SUMMARY

Up until the time of writing (August 1997) operational experience and generation levels have been good and in accordance with predictions.

R C Taylor
August 1997

Feasibility of energy recovery from a wastewater treatment scheme

M GRIFFIN MSc, CEng, MIMechE
Mott Ewbank Preece, Brighton, UK

SYNOPSIS

This paper describes a study that was undertaken to determine the potential of energy recovery from a new wastewater treatment scheme being built for the Kent towns of Dover and Folkestone. The paper reviews the potential sites and describes how their energy generating potential was evaluated using spreadsheet based software models. Two schemes demonstrated sufficient potential for further study. The paper describes potential operating problems that were envisaged and how the generating and control equipment was selected and arranged to achieve optimum performance at minimum cost. Designs for the schemes were developed such that costs and generating potential could be determined with sufficient accuracy to allow their financial rate of return to be estimated.

1 BACKGROUND

A new wastewater treatment scheme is currently being constructed for the Kent towns of Dover and Folkestone. For environmental reasons it was necessary to place the wastewater treatment works (WTW) away from the coast and the site selected was at Broomfield Bank, high on the South Downs between the two towns. Crude wastewater will be pumped to the treatment works before being discharged by gravity through a 3.3 km long sea water outfall at Dover. This paper describes a feasibility study that was conducted to examine the possibility of recovering some of the pumping costs by installing energy recovery turbines at suitable locations.

The work was divided into two stages. The first, pre-feasibility stage was essentially a screening process to determine whether it would be practicable to recover energy from the water flowing through the system. The more detailed feasibility stage developed the two most promising designs sufficiently to define capital and running costs and to determine the anticipated financial performance.

2   POTENTIAL SITES

The locations of the pipelines associated with the proposed water treatment system are shown on Figure 1, which also shows the elevations of the main features. In addition to the main pipeline routes to and from the treatment works a brief examination was made of the generating potential of the larger sewers within Dover and Folkestone. However, no sewers were identified with sufficient potential to warrant detailed investigation.

Two pipelines showed potential; the first was the final section of the wastewater transfer pipeline leading from Folkestone where it entered the WTW, whilst the second was the discharge pipeline from the WTW to the long sea outfall at Dover.

Table 1
Potential of Identified Sites

Site	Gross Head (m)	* Average Flow (l/s)	** Power Potential (kW)
Folkestone water transfer pipeline	43	132	55
Discharge Pipeline	68	427	284

* in 2001   ** Ignoring efficiency, losses etc.

In addition to the relatively low power potential of the Folkestone transfer pipeline, some of the effluent within it would be unscreened and it was considered that this might lead to blockage of the turbine water passages. These technical concerns, plus a poor financial performance led to this option being rejected at an early stage. However, the potential of the discharge pipeline was considered sufficient to warrant a more detailed investigation.

2.1 Discharge Pipeline

The route for the 3,383 m long discharge pipeline from Broomfield Bank to Dover has two distinct parts. The first section, which finishes within the playing fields of Dover College, is 1,475 m long and drops 28 m. This section was originally designed to run only partly full and would therefore remain unpressurised. The second section, which falls 40 m as it runs 1,906 m from the playing fields to the beginning of the long sea outfall, was designed as a pressure pipeline, although it would only be completely full at maximum flow. At lower flows the water surface would rise and fall in response to the tide level and the flow dependent head losses in the downstream pipework.

Two schemes were examined, the first utilised the head available in the lower section running from the playing fields to the outfall. The second took advantage of the full head available from Broomfield Bank; however, this scheme would have higher costs as the construction of the upper portion of the pipeline would have to be modified to become a full pressure pipeline.

3   POWER AND ENERGY MODELLING

Before investigating the selection of appropriate plant it was necessary to determine the power and energy that could be generated. A spreadsheet based model was developed which mirrored actual conditions as closely as possible. Using the flows given in Figure 2 the model calculated the head losses in the pipeline systems and hence the available net head. The power produced each hour was then determined from the net head, flow and estimates of machine efficiency. Hourly generation was summed to give daily and monthly energy.

3.1   Tide Effects

During initial model development the heads and flows available from the possible sites were not known and it was considered appropriate to include for the effect of the tide upon the available head. The first versions of the model therefore included hourly tide data (tables) for Dover over a three month period. An extract from a set of results is given in Table 2.

Table 2
Sample Results from Early Generation Model Including Tide Effects
Broomfield Bank to Dover Outfall

Date	Hourly Energy (kWh)			Daily Energy (kWh)	Tide Data (m above Ordnance datum)			Turbine Submergence (m)		
	Min	Avg	Max		Min	Avg	Max	Min	Avg	Max
02/01/95	160	177	193	4254	-3.1	0.0	3.1	3.5	4.7	8.3
03/01/95	157	177	194	4256	-3.2	0.0	3.2	3.5	4.7	8.0
03/01/95	155	177	194	4258	-3.1	0.0	3.2	3.5	4.6	7.8
08/02/95	149	178	209	4265	-1.7	0.1	1.8	3.5	3.9	5.2
09/02/95	149	178	206	4263	-1.4	0.1	1.5	3.5	3.9	5.1
10/02/95	150	178	204	4262	-1.4	0.1	1.5	3.5	4.0	4.9

The above results, extracted from daily figures for the first three months of 1995, are typical of the range that could occur. The tide range varies from over 6 m at the beginning of January to only 2.9 m at the start of February. The minimum daily energy generated during the three month period was 4,251 kWh with a maximum of 4,266 kWh, a variation of less than 0.5%. The small variation showed that for most cases the effect of the tide could be ignored.

The results in Table 2 also include the anticipated submergence of the turbine. These were calculated from the tide level and outfall head losses to determine the back pressure on the turbine. The results initially indicated that at times of low tide the turbine would have insufficient submergence. However, when the layout drawings of the outfall became available it was clear that there would be a high point 3.5 m above Ordnance Datum, where the pipeline crossed an existing structure. This feature ensured that the turbine would always be adequately submerged and explains the constant value in the results.

3.2 Plant Efficiency

Initially the model assumed that the overall generating plant efficiency remained constant at 75%. It was realised that this was towards the lower end of what might be expected but it was considered that a conservative approach was appropriate for these early screening investigations. When the overall parameters for the two sites had been determined, manufacturers were approached for budget prices and more exact estimates of plant performance. The data received included the variation of turbine and generator efficiency with output and this is shown in Table 3. The power and energy model was subsequently modified to calculate the overall generating efficiency for each hourly condition. The results showed a significant increase in the estimate of energy generated and demonstrated the benefit of including such effects into the model.

Table 3
Typical Variation of Plant Efficiency with Flow

Turbine Flow (l/s)	Turbine Efficiency (%)	Turbine Power (kW)	Generator Efficiency (%)	Overall Efficiency (%)
250	80.2	108	94.8	76.0
300	83.7	135	95.0	79.5
350	87	164	95.1	82.7
400	89.2	193	95.3	85.0
450	90.5	220	95.1	86.1
500	89.7	242	95.0	85.2

4   SELECTION OF GENERATING EQUIPMENT

As demonstrated by Figure 2 the flow into the treatment system is not constant with a minimum at night when activity is low before reaching a maximum in the early morning as people wake up. The flow then reduces before reaching another, somewhat lower peak, in the early evening. However, although it is not shown in Figure 2, the flow in the sewers increases markedly when it is raining as additional water enters the system. The generating plant had to be selected to cope with peak flows that were approximately double the normal, "dry weather" flows.

Due to the high relative velocities that occur within turbines, large concentrations of suspended solids would cause excessive wear. However, most untreated sewage is screened before it is pumped to Broomfield Bank and the treatment process includes settlement to remove heavier particles. Figure 3 shows the anticipated distribution of particulate matter in the treated effluent and it is clear that it is relatively benign with low levels of suspended solids with most of these of a non-abrasive nature. It was therefore decided to follow standard practice for the water treatment industry with internal components subject to high velocities manufactured from stainless steel. Other parts, such as pipework and turbine casings, would be made from mild steel or other lower cost materials.

4.1 Turbine

For maximum head, and thereby power and energy, the upstream water level should be maximised and the control system arranged to keep the inlet water level as high as possible. Due to the variation in the head losses in the outfall and the rise and fall of the tide there would be a relatively large variation in the water level downstream of the turbine. This is demonstrated by the data in Table 4.

Table 4
Variation in Tailwater Level with Flow and Tide Level

Outfall Head Losses at Min. Flow (m)	Outfall Head Losses at Max. Flow (m)	Minimum Tide Level (from Ordnance datum) (m)	Maximum Tide Level (above Ordnance datum) (m)	Maximum Tailwater Level (m)
3.6	19.8	-3.3	3.2	23.0

Small hydro turbines are often of the impulse type, such as the "Pelton" machine, as they are relatively simple and low cost devices. However, as they incorporate a jet that impinges on a runner rotating in air they have to have their outlet above the maximum downstream water level. The data in Table 4 demonstrates that an impulse machine would need to be placed at least 23 metres above Ordnance datum if it was to be able to run at all times and this would lead to a very significant loss of generated energy.

For maximum energy generation the turbine should preferably be of the reaction type as these run fully submerged and this would allow the turbine to utilise the maximum available net head at all times. The most common type of reaction turbine is the "Francis" machine and this can be manufactured to suit a range of available heads and power outputs. Table 5 contains the main parameters for the machines selected for the two locations examined

Table 5
Francis Turbine Design Parameters

Parameter	High Head Site (Broomfield Bank to Dover)	Low Head Site (Dover College to Dover)
Net Head	27 m	55 m
Best Efficiency Flow	500 l/s	500 l/s
Rotational Speed	1000 rev/min	1000 rev/min
Specific Speed	175	103
Runner Diameter	375 mm	365 mm

4.2 Turbine Bypass Valve

For the turbine to be taken out of service, either for maintenance or following a forced outage, bypass facilities would have to be provided. As discussed above, the maximum flow is approximately double the average flow. If the turbine were sized to pass the maximum flow then it would be running at significantly reduced efficiency when it was passing the average flow. If the bypass valve could be arranged to open at higher flows then the turbine capacity could be reduced to match its best efficiency flow to the average flow.

It was necessary to find a design of bypass valve which would have a very high level of reliability, to ensure that the flow through the outfall was rarely disrupted, whilst at the same time achieving accurate control so that the maximum possible flow could be passed through the turbine. The valve would also have to operate submerged with its discharge into the downstream pipe section. The Larner-Johnson type of valve was selected for the following reasons: -

- It is well proven and has been in service in similar situations for many years.
- The design is essentially self cleaning with a large flow control orifice that is unlikely to become blocked.
- The discharge flow from the valve is uniformly distributed and is therefore well suited to a pipeline installation.
- It can be powered by a range of control systems including directly by electric motors or by a hydraulic system.
- It can be arranged to be fail safe and open automatically following a fault.

The principle of operation of a typical Larner-Johnson valve is shown in Figure 4.

4.3   Generator and Associated Electrical Equipment

As there would be no requirements for the small generator to either run isolated or contribute to system frequency control, it could be of the induction type rather than a more expensive synchronous machine. No control of excitation is necessary with an induction generator and starting is straightforward with the machine being brought close to its running speed before being directly connected to the electrical system. Once the unit is generating its running speed will be limited by the grid frequency whilst the power output will simply match that generated by the turbine.

As discussed below, it was proposed to connect the generator into the local distribution system at the Elizabeth Street pumping works where a 2.5 MVA transformer was already installed to provide electrical supplies to existing pumping equipment. A spare electrical feeder was available on a switchboard to be installed as part of the revised water treatment system and this would allow the generator to be connected with little difficulty. The unit circuit breaker would be installed close to the generator with the connection to the pumping station switchboard by buried cable.

The rules controlling the technical requirements for connecting a generator to the national electricity supply are well defined and if the appropriate standard is called for, plant suppliers can provide equipment to suit the requirements without difficulty. However, correspondence with the local REC highlighted a potential problem, whereby the proposed induction generator might cause the night time voltage level (which is higher than normal due to lower system losses) to rise above the maximum permitted level. We studied this problem in some detail and concluded that by suitably setting the tapping ratio of the supply transformer, the voltage could be maintained such as to allow pumps to be started when supply voltages were low, whilst keeping below the maximum voltage limit when generating at night.

4.4 Control and Communications

It was a requirement that the generating plant should be fully automatic and need the minimum of intervention. A well established SCADA system is to be installed for remote monitoring of the water treatment facilities and it was agreed that the generating plant should be included within this overall scheme. The latest designs of programmable logic controllers (PLC) provide high levels of reliability at relatively low cost and include all the sequence control, calculation and alarm functions that might be required for a plant of this type. It was decided to base the control system around such equipment. To minimise cabling the PLC would be installed close to the plant rather than remotely within the pumping station. Contacts and cables would be provided to give status and alarm information to the SCADA system.

4.4.1 Flow and Level Control

Control of the upstream water level, which would also be the discharge point from the WTW, must be very reliable, as maloperation would risk spillage of raw sewage. Measurements of the turbine inlet pressure could be used to calculate the upstream water level if appropriate corrections for flow rate (derived form the turbine and bypass valve openings) were applied within the PLC. It was considered, however, that for safety reasons a signal connection would be necessary between Broomfield Bank and the turbine to open the bypass valve if the upstream water level became excessive. If communications facilities were to be provided for this facility, then the same system could be used to provide a more accurate direct measurement of the upstream water level. If the connection to Broomfield Bank should become lost then the turbine would continue to generate using local measurements generally as described above, although the target upstream level would necessarily be reduced to give an increased safety margin. The study determined that the most cost effective and reliable connection between the generating plant and Broomfield Bank would be by leased telephone lines.

5   PLANT LAYOUT

The site selected for the generating plant was where the discharge pipeline emerged from the cliff face in Dover, immediately below Archcliffe Fort. Unfortunately, being close to an historical monument meant that any buildings would be subject to stringent planning controls. Considerable time had already been expended in agreeing the original design in this area and to avoid any further delay changes had to be minimised.

The raw water pipeline from Dover leading to the WTW and the discharge pipeline from Broomfield Bank to the outfall, run in a common tunnel. Both pipelines will emerge from the cliff face below the local ground level before diverging and running underground to the pumping station and sea water outfall. The only indication of the presence of the tunnel would be a small retaining wall in the cliff face above the tunnel portal. Thrust blocks to react the forces from bends in the pipelines would also be buried. In order to minimise visual impact it was agreed that the generating plant would also have to be underground and the proposed arrangement in shown in Figure 5.

A Francis turbine introduces a 90° change in direction in flow and to save space and to minimise head losses the proposed plant room has been positioned so that the turbine coincided with a bend. The turbine bypass valve and connecting pipework have been positioned so that they form the hypotenuse of a triangle with the turbine forming the shorter two sides.

Although, as discussed in 4.2 above, the bypass valve should be very reliable, it would probably need to be removed periodically for maintenance. A separate bypass pipe was therefore included to allow the bypass valve to be removed. Except for the turbine inlet valve, which would have an powered actuator, opened and closed by the automatic sequencing equipment, it is proposed that the equipment isolation should use manually operated butterfly valves.

To minimise the need for access into the plant pit (where there would be the possibility of a build up of dangerous gases) it was proposed that the control equipment and generator circuit breaker should be installed within a small building erected within the retaining wall above the tunnel portal. Plant monitoring and electrical maintenance could therefore be relatively unrestricted. When necessary, personnel entry into the pit would be via a stairway, whilst equipment access would be from above via removable hatches, with heavy equipment lowered into the pit by mobile crane. A manually operated overhead crane would be provided for moving equipment within the plant room. Cables would be directly buried in the ground except at road crossings where they would run in ducts.

6   FINANCIAL ANALYSIS

The additional capital expenditure of including generating plant in the new wastewater treatment scheme will only be committed if a sufficiently high financial rate of return can be demonstrated. A financial model was developed which calculated the internal rate of returns that could be expected from potential schemes.

6.1   Costs

Preliminary designs and plant ratings were developed for all schemes that passed the initial screening. Manufacturers' budget prices were obtained for the main plant items such as turbines, generators and bypass valves. Estimates for the balance of plant and civil engineering works were made from in-house cost databases.

6.2   Income

When the study started in 1995 it was considered that it would be feasible to sell any power generated under the preferential terms provided by the Non Fossil Fuel Obligation (NFFO). The estimated income generated by such sales was therefore used for the financial calculations included in the pre-feasibility report. When the full feasibility studies were carried out in 1996 the situation had changed somewhat and it was then considered that hydro generation was less likely to be selected under NFFO. Alternative consumers of the power were therefore examined. It would be possible to enter into an arrangement to sell power directly to a local user but there has been little experience of such arrangements and it was considered to be beyond the scope of this study to estimate the terms that might be achieved. It was therefore decided to examine how the owners might consume the power themselves.

The generator would be connected to the electrical system at a pumping station which itself consumes a considerable amount of energy, currently purchased under a bulk supply arrangement. For most of a day, any power generated could be used within the pumping station and thereby reduce the amount purchased. However, at night both pumping and generating would probably be intermittent and there would be times when there would be generation but no significant consumption. Any excess generation would have to pass into the local distribution system. Although the arrangements could not be completely defined before the study was completed the base case for calculating income for the feasibility study was the value of reducing power purchased from others.

The cost of power purchased under commercial tariffs varies both with the time of day and the time of year. Table 6 shows in relative terms the variation in the value of power used in the analysis and this demonstrates the high cost of purchasing energy during the periods of maximum winter demand.

Table 6
Variation of Value of Power with Time of Day and Year

Time of day	Relative Cost of Power
00.30 to 07.00	0.40
08.00 to 00.30	1.00
16.00 to 19.00 (December to February inc.)	3.34

In addition to any income generated by reducing energy consumed there is an extra "TRIAD" charge levied upon consumers which depends upon the maximum demand during the winter months between 16.00 and 19.00 hours. Generation during this period would reduce the TRIAD charge.

The energy generating model discussed in Section 3 above was further developed to take account of the above effects as well as time related costs which should be deducted from the income. Figure 6 is a typical printout from the model for the scheme from Broomfield Bank to Dover. Although not included in this printout the average total energy that would be generated annually would be 1,818 MWh in the year 2001 rising to 2,010 MWh in 2021 when flow rates are expected to have increased.

6.3 Financial Performance

A simple spreadsheet based cash flow model was produced which calculated the expected internal rate of return and the net present value. Sensitivity analyses were included to show the effect of changes in capital cost, the power tariff and the amount of energy generated. Figure 7 shows the results corresponding to the energy model shown in Figure 6.

At the end of the study the anticipated internal rate of return for the base case project was 13.2%. Whilst this is not as high as that might normally be considered necessary to justify the additional capital expenditure, it does demonstrate that it would be possible to significantly reduce energy consumption and at the same time make a modest financial return. At the present time the study is being brought up to date following the completion of the detailed design of the WWTW and the outfall.

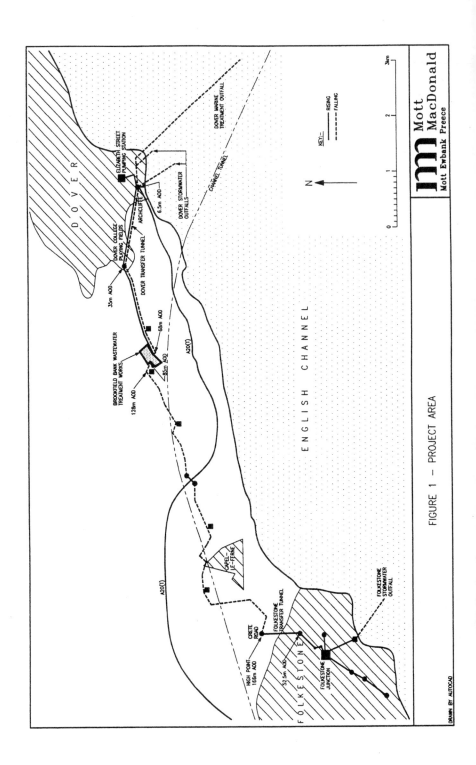

FIGURE 1 - PROJECT AREA

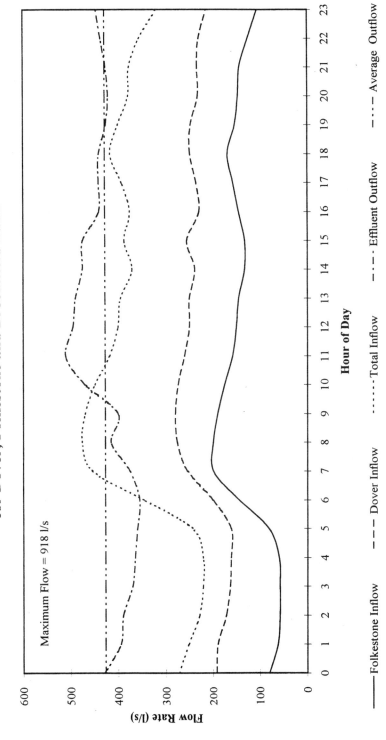

Figure 2: Typical Diurnal Variations in Flowrate During Dry Weather for Dover, Folkestone and Broomfield Bank

Maximum Flow = 918 l/s

Flow Rate (l/s)

Hour of Day

——— Folkestone Inflow   – – – Dover Inflow   · · · · · Total Inflow   – · – Effluent Outflow   – – – – Average Outflow

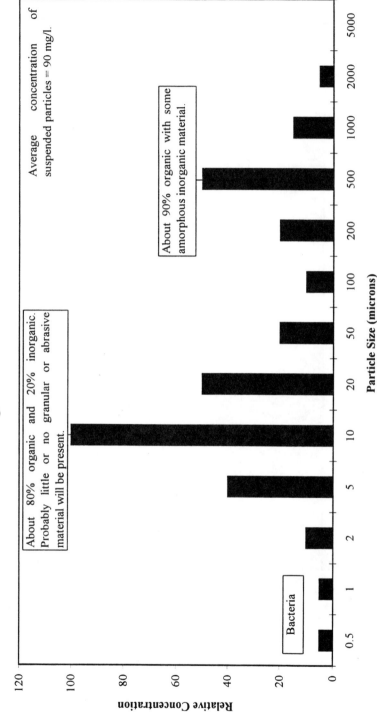

Figure 3: Approximate Particle Size Distribution in Settled Sewage Discharged from Broomfield Bank

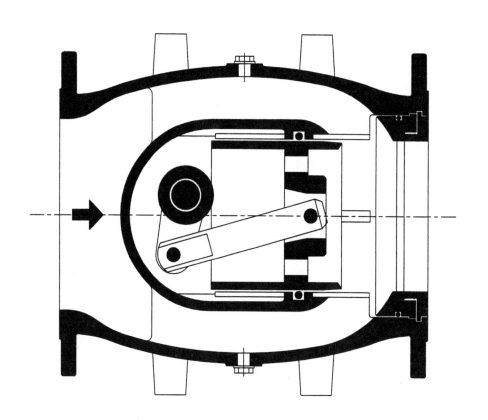

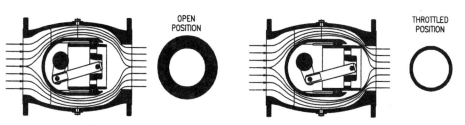

OPEN
POSITION

THROTTLED
POSITION

FLOW CROSS SECTIONS IN THE SEATING ZONE

FIGURE 4 – PRINCIPLE OF OPERATION OF
LARNER–JOHNSON VALVE

RAWN BY AUTOCAD

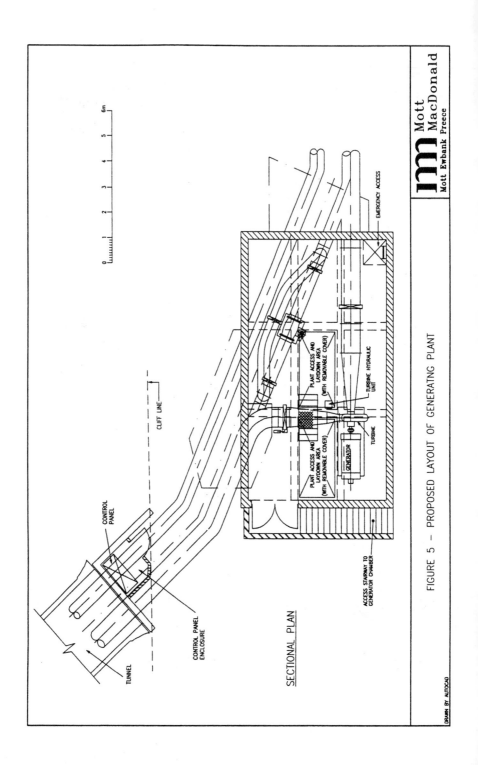

SECTIONAL PLAN

FIGURE 5 – PROPOSED LAYOUT OF GENERATING PLANT

Mott MacDonald
Mott Ewbank Preece

FIGURE 6 - TYPICAL DAILY OUTPUT FROM POWER AND ENERGY GENERATION MODEL
VARIABLE ENERGY VALUE THROUGH DAY AND YEAR
INCLUDING CABLE LOSSES AND TELECOM COSTS

	Pipe Dia (m)	Pipe Length (m)	Friction Factor	Bend etc. Losses (K)
WTW to Dover College Playing Field (1)	0.9	1475	0.01185	2.5
Dover College to Archcliffe Fort (2)	0.8	1514	0.01185	2
Archcliffe Fort to Marine Outfall (3)	0.8	392	0.01185	3.1
Marine Treatment Outfall (4)	0.7	1000	0.01185	N/A

Month 12

Average Daily Flow Rate 627.08 l/s

Design Top Water Level	68.00	m AOD
Turbine Centreline at Elevation	4.59	m AOD
Generator Efficiency	95.00%	
Zone Charge	£10.85	per kW
Line Impedance	0.0495	Ohms per km
Total Line Impedance	0.0149	Ohms
Cable Size	500	sqmm
Minimum Generation during TRIAD Perio	232	kW
Monthly Line Rental for Data Telephone Li	£28.00	(£84/quarter)

Time of Day (Hrs)	Percentage Of Hourly Flow Rate (l/s)	Flow Rate This Hour (l/s)	Head Loss Section (1) (mm)	Head Loss Section (2) (mm)	Head Loss Section (3) (mm)	Head Loss Section (4) (mm)	Total Head Loss (m)	Tide Elevation (m ACD)	Tide Elevation (m AOD)	Net Head for Generation (m)	Turbine Efficiency (%)	Energy Generated (kWh)	Energy Losses (kWh)	Energy Exported (kWh)	Value of Energy (p/kWh)	Income Generated (£)
00:00	100.8%	632	1.62	3.03	0.99	9.80	15.45	3.7	0.0	52.6	77.18	238.90	7.35	231.55	2.5967	6.01
01:00	92.6%	581	1.37	2.56	0.84	8.39	13.15	3.7	0.0	54.8	84.81	251.66	8.16	243.51	1.4860	3.62
02:00	91.4%	573	1.33	2.50	0.82	8.20	12.84	3.7	0.0	55.2	85.62	252.29	8.20	244.10	1.4860	3.63
03:00	86.7%	544	1.20	2.25	0.74	7.45	11.63	3.7	0.0	56.4	88.26	252.13	8.19	243.94	1.4860	3.63
04:00	85.5%	536	1.17	2.19	0.72	7.27	11.34	3.7	0.0	56.7	88.77	251.46	8.14	243.32	1.4860	3.62
05:00	84.4%	529	1.14	2.13	0.70	7.10	11.05	3.7	0.0	56.9	89.23	250.55	8.08	242.47	1.4860	3.60
06:00	83.2%	522	1.10	2.07	0.68	6.92	10.77	3.7	0.0	57.2	89.63	249.42	8.01	241.41	1.4860	3.59
07:00	87.9%	551	1.23	2.31	0.76	7.63	11.93	3.7	0.0	56.1	87.69	252.56	8.21	244.34	1.4860	3.63
08:00	97.3%	610	1.51	2.83	0.93	9.18	14.44	3.7	0.0	53.6	80.88	246.23	7.81	238.43	3.7073	8.84
09:00	93.8%	588	1.40	2.63	0.86	8.58	13.47	3.7	0.0	54.5	83.93	250.75	8.10	242.65	3.7073	9.00
10:00	110.2%	691	1.93	3.63	1.19	11.56	18.30	3.7	0.0	49.7	63.85	204.27	5.37	198.90	3.7073	7.37
11:00	119.5%	750	2.28	4.27	1.40	13.47	21.42	3.7	0.0	46.6	44.84	145.90	2.74	143.16	3.7073	5.31
12:00	116.0%	728	2.15	4.02	1.32	12.74	20.22	3.7	0.0	47.8	52.68	170.67	3.75	166.92	3.7073	6.19
13:00	114.8%	720	2.10	3.94	1.29	12.50	19.83	3.7	0.0	48.2	55.10	178.15	4.09	174.06	3.7073	6.45
14:00	111.3%	698	1.98	3.70	1.21	11.79	18.68	3.7	0.0	49.3	61.80	198.31	5.06	193.24	3.7073	7.16
15:00	111.3%	698	1.98	3.70	1.21	11.79	18.68	3.7	0.0	49.3	61.80	198.31	5.06	193.24	3.7073	7.16
16:00	103.1%	647	1.70	3.18	1.04	10.22	16.14	3.7	0.0	51.9	74.34	232.35	6.95	225.40	16.1326	36.36
17:00	103.1%	647	1.70	3.18	1.04	10.22	16.14	3.7	0.0	51.9	74.34	232.35	6.95	225.40	16.1326	36.36
18:00	103.1%	647	1.70	3.18	1.04	10.22	16.14	3.7	0.0	51.9	74.34	232.35	6.95	225.40	16.1326	36.36
19:00	99.6%	625	1.58	2.96	0.97	9.59	15.11	3.7	0.0	52.9	78.49	241.67	7.52	234.15	3.7073	8.68
20:00	98.4%	617	1.54	2.89	0.95	9.38	14.77	3.7	0.0	53.2	79.72	244.11	7.67	236.44	3.7073	8.77
21:00	99.6%	625	1.58	2.96	0.97	9.59	15.11	3.7	0.0	52.9	78.49	241.67	7.52	234.15	3.7073	8.68
22:00	102.0%	639	1.66	3.11	1.02	10.01	15.79	3.7	0.0	52.2	75.80	235.79	7.16	228.63	3.7073	8.48
23:00	104.3%	654	1.73	3.25	1.06	10.44	16.49	3.7	0.0	51.5	72.80	228.56	6.73	221.84	3.7073	8.22
Daily Totals											Energy (kWh)	5,480	164	5,317	Income (£)	241
Monthly Totals											Energy (kWh)	169,892	5,077	164,815	Income (£)	7,462

FIGURE 7 - TYPICAL RATE OF RETURN CALCULATION WITH SENSITIVITY EFFECTS

Input Data

Capital Cost of Plant (from Figure 6.1)		£196,989
Total Capital Cost including Civil Works		£444,492
Capital Cost With Uprated Cable		£453,932
Annual O&M and Connection Costs		
(Proportion of Plant Capital Cost)		1.5%
"Average" Energy in 2001	1.808 (GWh)	"Average" Income (2001) £64,395
"Average" Energy in 2021	2.004 (GWh)	"Average" Income (2021) £70,960
Annual Discount Rate (NPV)		10.00%
Cable Size		400 sqmm

Results

Nett Present Value of Net Cash Flow
at Annual Discount rate   £109,114   Internal Rate of Return of Net Cash Flow   13.23%

Year	Capital Expenditure (£)	O&M & other Regular Costs (£)	Total Outgoings (£)	Annual Energy Generated (GWh)	Income from Power Sales (£)	Net Cash Flow (£)
1998	453,932	0	453,932	0.000	0	(453,932)
1999	0	2,955	2,955	1.788	63,739	60,784
2000	0	2,955	2,955	1.798	64,067	61,112
2001	0	2,955	2,955	1.808	64,395	61,440
		The years 2002 to 2019 have been left out for clarity				
2020	0	2,955	2,955	1.994	70,632	67,677
2021	0	2,955	2,955	2.004	70,960	68,005
2022	0	2,955	2,955	2.014	71,288	68,333
2023	0	2,955	2,955	2.023	71,617	68,662

EFFECT OF ANNUAL ENERGY
1996 Base Tariff £ 437,092 Capital Cost

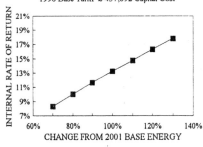

CHANGE FROM 2001 BASE ENERGY

EFFECT OF VALUE OF ENERGY
£437,092 Capital Cost, 1,824 GWh per year (2001)

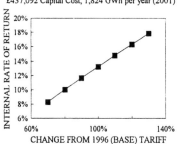

CHANGE FROM 1996 (BASE) TARIFF

EFFECT OF CAPITAL COST
1996 Base Tariff, 1.82·GWh per Year (2001)

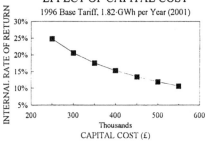

CAPITAL COST (£)

he Pergau hydro-electric project – a project overview

PATEL BSc, CEng, MICE, MIStructE
lfour Beatty International Limited, Sidcup, UK
1 BAIN CEng, FIMechE, FIMarE
dro Power Engineering Limited, Orpington, UK

INTRODUCTION

The Pergau Hydro Electric Project was constructed as a design and construct contract in Northern Peninsular Malaysia by an Anglo-Malaysian Joint Venture. Final commissioning of the generating plant was completed in August 1997.

This paper will describe the Project, which was the largest hydro electric development undertaken by UK companies in recent years.

It is a 600MW installation of four 150MW Francis turbine generating sets operating at 495m net head. The powerhouse is underground and houses both the generating units and the generator transformers.

The main features of the project are the 75m high earth fill dam, power tunnels leading to an underground power cavern and a reregulating pond at the exit of the tailrace tunnel.

Additional features are a pumping station on the Terang River which houses five variable speed pumps capable of delivering 10m3/s at 68m head. Water is pumped along a 24km long aqueduct tunnel, which also collects water from six small intakes before discharging into the Kuala Yong Dam reservoir.

PROJECT OVERVIEW

The project is located in northern peninsular Malaysia approximately 100km west of the Kelantan state capital of Khota Baru.

See Figure 1.

Outline of Project

The central feature of the project is the Kuala Yong Dam and Spillway. The power intake diverts water from the reservoir through a single pressure headrace tunnel to the surge shaft at the top of a single inclined pressure shaft. Bifurcations at the underground powerhouse feed the high pressure water to the four vertical axis turbine generating units. The turbines discharge into a single tailrace tunnel into a reregulating pond. The reregulating pond includes a Siphon Spillway and Gated Outlet Structure discharging to the River Pergau.

The reservoir is fed by the River Pergau and augmented by a Pumping Station at the River Terang which pumps water through an aqueduct tunnel which is also fed by six small intakes.

See Figures 2 and 3.

The Terang Pumping Station and Weir

As the River Terang intake is below the level of the aqueduct tunnel and its hydraulic line, the water has to be pumped up to the tunnel. The pumping station is located on the River Terang 3 immediately downstream of the confluence of Rivers Terang 1 and 2. A small headpond created by a concrete faced rockfill weir is located on the left bank and consists of a log deflector and skimmer wall, five screened intakes of 2.5m x 3.75m, five desilting basins 20m x 8m and a pumphouse 29m x 27m x 22m high. The height and crest length of the weir is 8m x 32m respectively.

The flow to each pump forebay passes through a bellmouth inlet into the suction pipework within the pumphouse. The pumphouse is equipped with five identical vertical axis, variable speed pumps. Each pump suction being connected to a dedicated forebay and discharges into a common steel manifold. The total station discharge passes through a 2.4m diameter guard valve before entering the aqueduct tunnel.

Electrical power to the pumping station is supplied from Pergau Hydro Electric Power Station from a 33kV double circuit overhead transmission line.

See Photos 1 and 2.

The Aqueduct System

The aqueduct system transfers water from the Terang Pumping Station and six minor intakes (via dropshafts) to the Kuala Yong reservoir. The first 150m of the aqueduct tunnel from the Terang outlet is steel lined. A surge shaft (8m diameter, 55m depth) is incorporated into the system to suppress the surge and transient water pressures that develop during pump start up and shutdowns. The remainder of the aqueduct is unlined (except for primary support or as a secondary lining for structural reasons) with a concrete invert. The total length of the aqueduct is 24km with a cross sectional area of approximately 10.3m2. The total design flow through the aqueduct is 17.05m3/s. (9.8m3/s is from the Terang Pumping Station).

The six minor intakes from the small rivers in the catchment area Suda, Renyok 2, Renyok 3, Long 1 and Long 2 supplement the water system. Each intake has dropshafts ranging from 43m to 134m depths. The Long 1 and Renyok 3 dropshafts have finished diameters of 2.5m while the others have diameters of 1.75m. All dropshafts are concrete lined. Each intake includes an overflow spillway, intake structure with gates and screens, desanders and outlet gates and bypass and isolating gates. All dropshafts are concrete lined. The dropshaft excavation was by raise boring to a diameter of 3m.

Two maintenance adits have been provided to allow for four wheel vehicle access into the aqueduct. The Renyok maintenance adit is 430m in length and the Long adit 450m. The Long maintenance adit also acts as an aqueduct dewatering channel. An 800mm diameter butterfly valve is utilised to control dewatering flow.

See Figures 2 and 3.

The Kuala Yong Dam and Spillway

The dam, which is situated on the River Pergau, is a zoned earth dam with a central core founded on weathered granitic rocks. The core/foundation contact on the left abutment is in the form of a cut-off trench, which extends to fresh rock. On the right abutment weathered rock has been left in place and a jet grouted cut-off in combination with a drainage curtain retained for seepage control. The pressure grout curtain is installed beneath the centre of the core for the full length of the cut-off trench on the left abutment and in the centre of the dam area and as an extension beneath the jet cut-off as far as the spillway on the right abutment.

The maximum height of the main dam is 75m with a crest level of 750m. The reservoir has a live storage capacity of 53.5 million m3 (gross storage capacity is 62.5 million m3).

The free overflow concrete lined chute spillway (with terminal flip bucket) discharges flood inflows, which exceed the regulating capacity of the reservoir and/or discharges normal river flows during periods of prolonged station outages. The crest length of the spillway is 149m long with a trough 48m long, control section

20m long and chute 220m long. The flip bucket at the end of the chute has a radius of 15m and is 40m above the plunge pool.

A Riparian outlet provides compensation flows from the reservoir at the base of the dam and allows for controlled downstream flows. A 500mm diameter riparian valve is installed.

See Photo 3 and Figure 3.

The Power Tunnels and Shafts

The power tunnels and shafts convey the flow of water from the intake structure at the reservoir to the underground powerhouse complex and from the turbine outlets to the tailrace tunnel and the reregulating pond.

The tunnels and shafts consist mainly of:

- **Intake Shaft**

A 27m high vertical shaft suspended from the rear of the intake structure base slab. The internal profile varies from 5.3m square to 6m diameter.

- **Low Pressure Headrace Tunnel**

A 6m internal diameter concrete lined tunnel 1001m long extending from the bottom from the bottom of the intake shaft to the pressure shaft junction.

- **Surge and Riser Shaft**

An 11m internal diameter concrete lined 63m vertical shaft extending from ground level to the riser shaft. It is topped by a 25m x 25m expansion chamber with 3m high walls. The riser shaft is 8m internal diameter concrete lined 28m vertical shaft extending between the bottom of the surge shaft and low pressure tunnel.

- **Inclined Pressure Shaft**

A 472m long inclined 60 degree shaft extending from the end of the low pressure headrace tunnel to the inlet of the high pressure tunnel. The internal diameter of the shaft varies from 6m at the concrete lined inlet to 5m at the steel lined outlet.

- **High Pressure Tunnel**

A 275 m long steel lined tunnel extending from the bottom of the inclined pressure shaft to the underground turbine inlet bifurcations. The diameters vary from 5m at the pressure shaft to 1.8m at the turbine inlets.

From the turbine draft tube outlet discharges, the downstream manifolds and tunnels are :

- **Draft Tube Tunnel**

Four (one for each turbine) 4.5m diameter inclined concrete lined each 16.m long. Connecting the turbine draft tubes to the tailrace tunnel manifold.

- **Tailrace Manifold**

An 8m diameter concrete lined manifold 56.4m long. The discharge downstream is to the tailrace tunnel and surge shaft.

- **Tailrace Surge Shaft**

A 7m internal diameter 305m long concrete lined shaft. Connected to the crown of the tailrace manifold by a 2m long shaft. This shaft is used to contain hydraulic surges in the 2.4km tailrace tunnel during turbine load rejections and rapid onloading.

- **Tailrace Tunnel**

A 6m internal diameter 2780m long concrete lined tunnel. The tunnel extends from the station manifold to the tailrace outfall at the reregulating pond.

See Figure 3 for complete system.

The Cavern Powerhouse

The cavern powerhouse houses all the generating and electrical and mechanical auxiliary plant.

The cavern is 30m wide, 96m long and 37m high.

Four vertical Francis turbine generating units are installed together with the turbine main inlet valves. Auxiliary plant includes the primary and secondary cooling water and heat exchanger systems and oil handling plant. The turbine draft tube gates are also located within the powerhouse cavern together with the hydraulic operating equipment.

The powerhouse includes a loading bay, maintenance workshop and stores at one end and a control room, battery room and ventilation plant room at the other end.

Adjacent to the powerhouse cavern is a transformer hall, which houses the main generator transformers. This hall is 80m long and 15m wide.

Access to the cavern is by a main access tunnel which is 1052m long, 8m wide and 7m high (horseshoe shaped). There is a separate tunnel serving two purposes, one for ventilation and the other for routing the power and control cables to the surface switchyard and control building. This tunnel is780m long, 7m high and 8m wide (horseshoe shape).

See Photos 4 and 5 and Figure 4.

Reregulating Pond

An unusual but important feature of this project is a reregulating pond at the outlet of the tailrace tunnel.

The regulating pond contains the power generation peak flow and allows controlled flows into the River Pergau to avoid damage to the existing river bed and flood conditions downstream. It can regulate the downstream river discharges so that near constant river flows can be maintained.

The pond embankment consists of broadly graded granular fill material with an impermeable HDPE geomembrane. Embankment volume is 770,000m3. The embankment has an internal slope o 1:2.9 and an external slope of 1:19. The crest elevation is at 124.5m and the invert level at 108.6m. Pond storage capacity is 1,520,000m3.

The outlet structure consists of a twin 3m high x 4m wide culvert. Radial gates are provided to release the outflow downstream at a controlled rate. An emergency siphon spillway capable discharging flows in excess of 141m3/s is provided.

See Photo 6.

Chilled Ventilation Plant

It is usual to provide ventilation for an underground station. For Pergau unusually high temperatures are experienced in the cavern due to the geological conditions (hot rocks) and the existence of hot springs. The ventilation plant installed included a surface chiller plant which provides chilled water to the ventilation plant installed in the powerhouse cavern.

The powerhouse cavern is supplied with cooling air ventilation and extract ventilation to maintain the specified ambient temperatures (25oC).

Fresh air is drawn down the main access tunnel at 25m3/s. From the recirculation air tunnel 10m3/s of air is added to give an overall air supply of 35m3/s. Within the dampered filtered/cooling mixing chamber, located in the powerhouse roof adit, a further 25m3/s of air is added from a high level bell mouth return air duct.

The air is circulated throughout the cavern by two of three main centrifugal fans sets.

Extract air is drawn through from the main cavern and through the transformer enclosures into the transformer hall ductwork and onto the main axial extract fans. The air discharges through the cable and ventilation tunnel out to atmosphere at the tunnel portal and surface.

Control and Monitoring

The control and monitoring at the Pergau Hydro Electric Power Station and its associated outstations including the Terang Pumping Station is carried out using a microprocessor based Distributed Control System (DCS).

Control rooms are located in the powerhouse cavern and the surface control room and separately at Terang Pumping Station. It is not the intention of this paper to discuss the details of the systems installed.

Acknowledgements

The authors acknowledge with thanks the permission of Tenaga Nasional Berhad (TNB) of Malaysia and the Joint Venture Kerjaya Balfour Beatty Cementation Sdn.Bhd. for permission to present this paper.

MAIN TECHNICAL DATA AND INFORMATION

Kuala Yong Dam and Reservoir

Crest Elevation	EL 642.00m
Maximum Dam Height	75m
Crest Length	750m
Full Supply Level	EL 636.00m
Minimum Operating Level	EL 615.00m
Live Storage Capacity	53,500,000 m3
Total Excavation of Rock and Soil	4,700,000 m3
Total fill materials	2,275,000 m3

Reregulating Pond

Crest Elevation	EL 124.50m
Invert Level	EL 108.60m
Pond storage Volume	1,510,000 m3
Restricted Maximum Outflow	68m3/s

General Civil Works

Total Excavations for Tunnels, Shafts and Caverns	700,000m3
Total Concrete Batched	390,000m3

The Aqueduct System

Total Length of the Aqueduct Tunnel	24km
Total Design Flow (includes 9.8m3/s from Terang)	17.05m3/s

Power and Mechanical Plant

In the Powerhouse
Four Units

Vertical Francis Turbines	153MW rated at 495m net head
Generators	166.7MVA, 150MW, 16kV 428.57rpm, 50Hz
Generator Transformers	275/16.9kv, 180MVA
Distributed Control System	with inbuilt dual redundancy

Terang Pumping Station
Five Sets

Vertical Split Casing Pumps	1.96m3/s,68m head
	955rpm to 643rpm
Vertical Flange Mounted Motors	1705kW at 985rpm
Totally Enclosed CACA, Induction	

Vertical Variable Speed Turbo Coupling Maximum 955rpm

At Power Intake

Bulkhead Gate, Slide Double Leaf	5.5m high x 2.5m wide
Three gates	
Guard Gates, Double Leaf Fixed Wheel	
With hydraulic operating cylinder	5.5m high x 2.5m wide

Electrical Power Transmission

275kv Line	2.5km
33kV Line	27km
11kV Line	7.5km

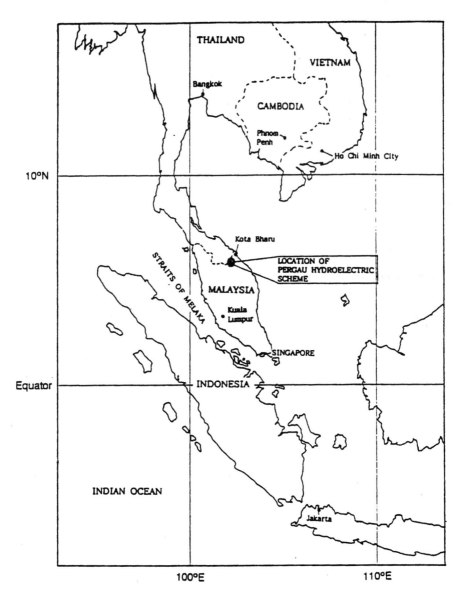

Figure 1 Location sketch

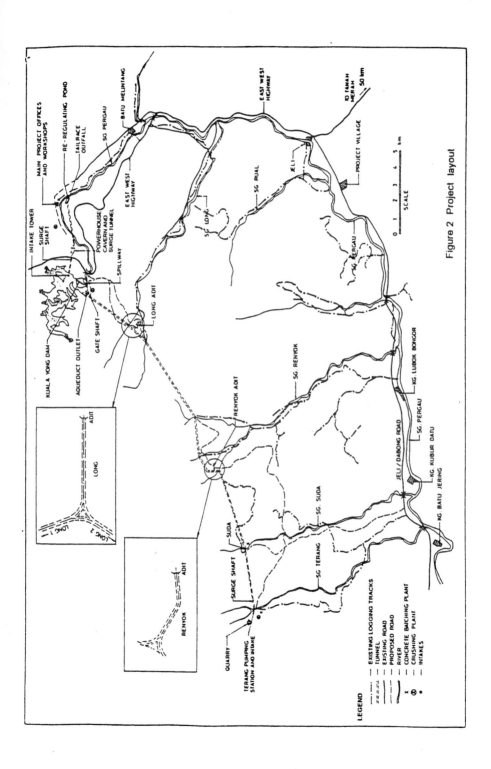

Figure 2 Project layout

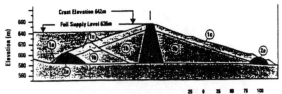

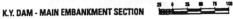

KEY	
1a	Embankment core fill
1b	Inner shell zone
1c	Outer shell zone
2a	Abutment stabilisation

K.Y. DAM - MAIN EMBANKMENT SECTION

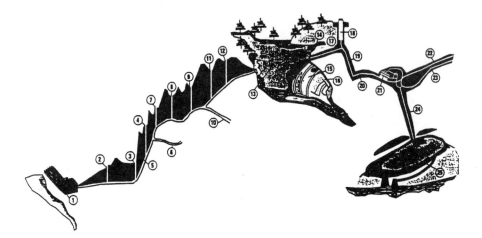

KEY			
1	Terang Pumping Station	14	Power Intake Structure
2	Aqueduct Surge Shaft	15	Kuala Yong Dam & Spillway
3	Suda Intake	16	Diversion Tunnel
4	Renyok 3 Intake	17	Low Pressure Headrace Tunnel
5	Aqueduct Tunnel	18	Surge Shaft
6	Renyok Maintenance Adit	19	Incline Pressure Shaft
7	Renyok 2 Intake	20	High Pressure Headrace Tunnel
8	Renyok 1 Intake	21	Powerhouse Complex
9	Long 2 Intake	22	Cable & Ventilation Tunnel
10	Long Maintenance Adit	23	Main Access Tunnel
11	Long 1 Intake	24	Tailrace Tunnel
12	Gate Shaft	25	Reregulating Pond With Siphon
13	Reservoir		Spillway & Outlet Structure

Figure 3 Project key elements.

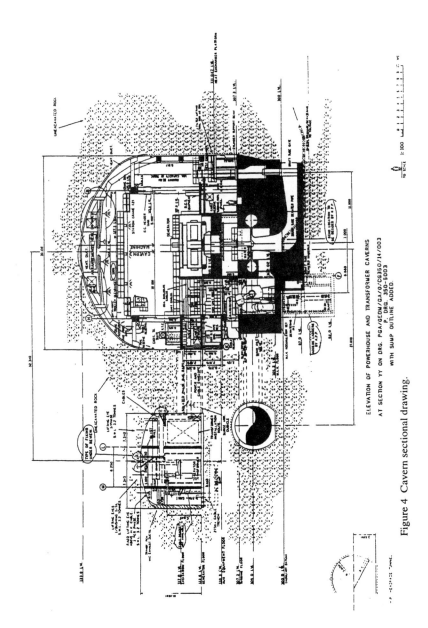

ELEVATION OF POWERHOUSE AND TRANSFORMER CAVERNS
AT SECTION YY ON DRG. PGA/GEDM/GA/O/CG350/14/003
P. DRG. 350-5003
WITH SUMP OUTLINE ADDED.

Figure 4 Cavern sectional drawing.

Photo 1 Terang Pumping Station.

Photo 2 Terang Pumping Station.

Photo 3 Kuala Yong Dam and Spillway

Balfour Beatty

Pergau River Hydro-electric Scheme, Malaysia

Balfour Beatty led the Joint Venture for this major turnkey contract and was instrumental in securing finance for the project in the mountainous north of the Malaysia Peninsular close to the Thailand border in Kelantan province.

Contract period:
66 months
Client:
Malaysian Electricity Board
Consultants:
Snowy Mountains Engineering Corporation

BICCGroup

Photo 4 Powerhouse construction.

Photo 5 Powerhouse complete.

Photo 6 Re-regulating pond.

Review of the hydraulic high-pressure components at Dinorwig and Ffestiniog power stations

P MORGAN BSc, MEng
First Hydro Company, Dinorwig Power Station, Llanberis, UK

SYNOPSIS

Since early operation both Dinorwig and Ffestiniog Pumped Storage Power Stations in North Wales, have both provided an excellent service to the UK grid system, being called upon to provide many services to facilitate the operation of the transmission system.

It is from this operation of pumped storage plant, the high pressure components are subjected daily, to a considerable number of mode changes. These mode changes are required to provide the customer with a service which will involve starting the pumping and generating units several times each day. This running regime puts a very onerous condition on the plant components especially the spiral and the high pressure penstock.

This paper looks at the pressure system to give an assurance that the system is fit for purpose to continue its operating life for a further 25 years for both power plants.

HIGH PRESSURE COMPONENTS

The high pressure components may be defined as those that are subjected to full penstock pressure, that is 60 bar at Dinorwig and 31 and 33 bar respectively for turbine and pump at Ffestiniog Power Station. These are recognised as the A section penstock, intermediate penstock, turbine main inlet valve, pump/turbine spirals and man access points, seen in Fig 1.

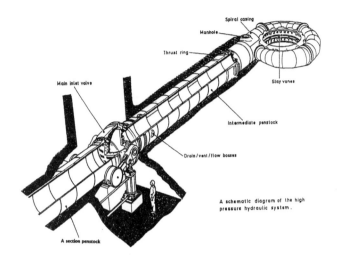

Fig 1

DINORWIG POWER STATION

The six reversible pump/turbine units at Dinorwig Power Station are situated on the outskirts of Caernarfon in Gwynedd, North Wales, are each rated at 288MW and transfer water between Peris and Marchlyn Lakes.

During commissioning in 1981, Dinorwig high pressure components were subject to a 100% ultrasonic Non-Destructive testing examination by automated pre-service inspection and manual in-service inspection followed by an assessment of high pressure components by fracture mechanics (1, 2, 3 and 4).

From these records and reports on non destructive testing, the long term inspection and testing philosophy could be defined based on the number of spiral pressurisation's.

Since commissioning in 1983 Dinorwig's spiral casings have been subjected to the following number of pressurisation's.

Unit No.	Total No. of pressurisation's to date Pump/Turbine spiral
1	81845
2	71572
3	83181
4	74259
5	69890
6	64277

FFESTINIOG POWER STATION

During 1965 the first pumped storage power station in the United Kingdom was commissioned, at Blaenau Ffestiniog in North Wales, comprising of four Francis designed turbines rated at 75MW and four twin suction pumps also rated at 75MW. These units were installed to transfer water between Tan-y-Grisiau and Stwlan Lake at a net head of 295m. This equates to a working pressure of 30 bar. The turbines were later rerated to 90MW, the pump and turbines are mechanically connected by a hydraulically operated coupling.

Since commissioning Ffestiniog in 1965 each spiral casing has been subjected to an extimated 116,000 pressurisations.

PHILOSOPHY OF HIGH PRESSURE COMPONENTS FOR DINORWIG AND FFESTINIOG

The following stages were identified as being essential for the review:- (see Fig 2)

Stage I

Before any testing programme can start all the life influencing factors external to the high pressure components need to be identified these include:-

1. Water quality
2. Load history to date
3. Historical data
4. Further life expectancy
5. Future loading expectancy.

During stage 1 drawings need to be made available or produced that identify all the welds associated with the high pressure components to be examined.

Stage II

Stage II may be defined as the non destructive testing for volumetric and surface breaking defects of the fabricated joints identified in Stage I above. The extent of

REVIEW OF HYDRAULIC PRESSURE COMPONENTS

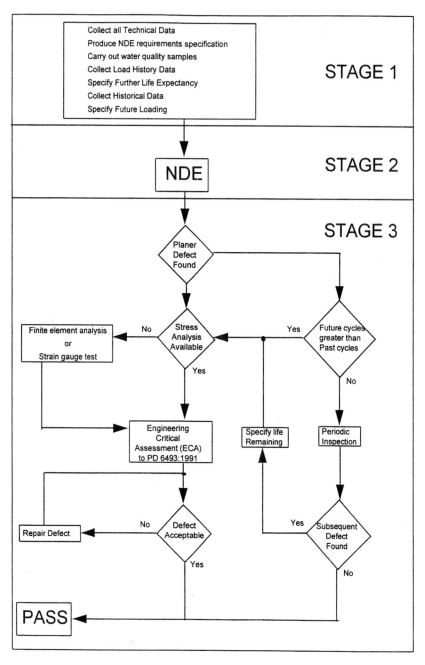

Fig 2

examination carried out would be governed by the outage time available to carry out the work.

First Hydro Company employed contractors for paint removal, and for non destructive testing by ultrasonic and magnetic particles. All reportable indications were accurately sized by time of flight non destructive testing techniques.

Stage III

Following non destructive testing, the results were given to a designer for assessment, this involved preparing a finite element model of the pump and turbine inlet valves, the pump and turbine spirals so as to determine the high stress areas.

Because of the complexity of the finite element model, the contractor was requested, to verify by hand calculation, certain predictions made by the model.

Stage III is therefore, the assessment of defects to PD6493: 1991 and to predict the number of cycles/pressurisation's available before the next inspection is due. Indications that show no change in length between periods of inspection, may be considered to be of less risk.

REVIEW OF HIGH PRESSURE COMPONENTS

Dinorwig

Since the design, construction and commissioning of Dinorwig, stages I, II and III referred to above, have been implemented. With the effect of water (5) on the fatigue life of manganese steels and loading of the units had been considered well before manufacture had even started.

First Hydro Company are therefore in the process of implementing the recommendations, as specified in the reappraisal for inspection specified in 1988 by the North West Region of the then CEGB (6), see Table 1.

REVISED INSPECTION SCHEDULE

AREA TO BE INSPECTED	PREVIOUS INSPECTION FREQUENCY		REVISED INSPECTION FREQUENCY	
	SELECTIVE	FULL	SELECTIVE	FULL
Upper Penstock (Embedded)	N/A	20 yrs	If penstock dewatered	None
Upper Penstock (In MIV Gallery)	N/A	20 yrs	27 yrs	None
MIV	N/A	None	None	None
Intermediate Penstock Welds	N/A	6 yrs	18 yrs	None
Boss Castings	N/A	U/S & MPI 18 yrs	MPI only 6 yrs	None
Manhole Castings	N/A	U/S & MPI 2 yrs	MPI only 6 yrs	None None
Spiral Casing Welds	N/A	2 yrs	6 yrs	18 yrs
Stay Vane Castings	N/A	6 yrs	Cavitated Areas 6 yrs	None None
Stay Ring Casting Local to Stay Ring Weld and Partition Stay	N/A	None	6 yrs	ASAP & 18 yrs
Manhole Door Bolts	N/A	None	None	MPI ASAP & 6 yrs

Table 1

Ffestiniog

The problems associated with the hydraulic testing of turbine spirals at works is well documented (7) the original turbine spiral for unit No. 4 failed catastrophically at works, slightly below maximum test pressure.

In view of this, the manufacturer decided to carry out ultrasonic examination of all 4 turbine spiral casings. For units 1 and 2 defects were found in both casings, along the spiral to speed ring weld, adjacent to plate No. 4. Following assessment of the indication, by fracture mechanics by the CEGB North West Region, Scientific Services in 1977, both the turbine spirals at the above locations have been subjected to periodic testing using ultrasonic so as to monitor for any change in defect length (8).

It is from this limited available information, that the review for Ffestiniog has been undertaken over the last 18 months, to cover all aspects of inspection on the pressure side of the components.

NDT TECHNIQUES FOR INSPECTION OF PUMP/TURBINE SPIRALS

Preparation

During ultrasonic examination for volumetric defects and magnetic particle examination for surface defects, the welds need to be identified, paintwork removed, and the surface polished to a minimum of 125 cla. for a band width each side of the weld centre line of 100mm.

Testing

Testing shall then proceed to agreed testing and reporting techniques and criteria. First Hydro Company have adopted a reporting criteria of 3mm for ultrasonic examination with its contractors.

The time of flight technique being used so as to size accurately the defect length. Surface breaking defects were reported as found on check sheets and were examined for depth using ultrasonic examination probes.

During the 1996 examination of Unit No. 4 at Ffestiniog a surface breaking defect having a total length of 200mm was found on the turbine spiral stayring to plate weld.

Metallurgical Assessment

Two areas of the above defect were metallurgically prepared and polished and the defect tips replicated. Laboratory examination, using an incident light microscope, revealed that the defects were intergranular, discontinuous and in most part oxide filled. They were caused by hydrogen cracking during the welding process and determined to be dormant. The unit was therefore returned to normal service on completion of its' planned outage.

Evaluation of alternative techniques for non destructive testing

It is interesting to note that because of the problems associated with reliability of the multi-head auto scanning unit, First Hydro Company have abandoned its use for non destructive testing purposes.

During the outage on Units 1 and 2 at Ffestiniog in 1997, it is proposed that the same inspection procedure be adopted for the inspection of the pump and turbine spirals as was used in 1996.

Paint removal, surface preparation, followed by cleaning of the couplant and repainting at Ffestiniog and Dinorwig pump and turbine spirals can be time consuming and costly. When repainting, the break in the joint between old and the new, means that the paint to steel bond is not as good as the original.

At Dinorwig because of the cost and time involved in removing the paintwork in the pump turbine spirals, First Hydro Company are in the process of carrying out feasibility tests so as to be able to carry out non destructive testing examinations without the need for paint removal.

First, so as to identify location of the welds, by not removing the paint work, eddy current probes which look for the change in material permability between the weld material and the parent metal, are being evaluated. This technique is also capable of detecting surface breaking defects.

Examination for volumetric defects, through the paint work is then carried out by time of flight.

Following the manufacture of a test block which has a series of defects introduced into the weld, both faces were painted to a thickness of 300 microns of proprietary turbine spiral paint.

The test block was then made available to various inspection companies providing the time of flight service, so to demonstrate their ability in identifying the defects through the paint work.

THE WAY FORWARD

Dinorwig

During 1981 - 1982 pre-service inspection, a total of 1731 indications were identified. Of these, only 39 were estimated to have a through wall thickness of greater than 3mm. Only 48 indications required further non destructive testing to determine the precise nature of the reflector. In all but one case, the indications could be resolved into a number of discrete reflectors, each classified as acceptable against the relevant fracture mechanics standard (9). The indication was revisited in Unit No. 1 circumferential weld R4 in 1985, 1987 and 1993 and was confirmed as being of 140mm in length, with maximum through wall extent less than 3mm.

Ffestiniog

The retrieval of any history on the extent of any testing or inspection of the Ffestiniog high pressure components has proven to be limited, and become more difficult with the loss of the North West Region, Scientific Service during the break up of the CEGB in the late eighties. This responsibility, analysis and interpretation now comes under the Engineering Department, First Hydro Company and it is from this philosophy that we are following a similar policy for that of Dinorwig and hope to give new inspection and monitoring periods to satisfy the protection and long operation of plant.

During 1995, First Hydro Company, in reconsidering its safety case for continued operation of Ffestiniog power station, decided to embark on a process of assessing the integrity of the water system high pressure components at Ffestiniog.

To date longitudinal and circumferential welds associated with two turbine and pump spirals have been examined for defects.

Finite element models have been prepared for both spirals so as to identify the stresses at the various locations.

From the stress levels identified in the finite element model, fracture mechanics assessment calculations are being prepared so as to determine the life expectancy and inspection frequency for the defect location.

CONCLUSION

It can be seen from the paper, a clear policy has been derived for Dinorwig and a revised frequency of inspection of all the pressure parts has been implemented. In Ffestiniog this policy is being developed, both with the FE modelling of the HP Pressure components, and the full plant NDT testing that has been carried out on two pump/turbine high pressure components during the outages of 1997. It is from this information and new NDT techniques that the full finger printing will be checked against the modelling and a similar philosophy will be derived at Ffestiniog.

References

1. Fracture mechanics assessment of the Main Inlet Valve Body terminal pipe and flange. Report No. NW/SSD/SR/158/81.

2. Fracture mechanics assessment of the pump turbine. Report No. NW/SSD/SR/159/81.

3. Fracture mechanics assessment of the High Pressure A - Section tunnels linings and intermediate penstock. Report No. NW/SSD/SR/160/81.

4. Fracture mechanics assessment of the Bosses and Manholes. Report No. NWR/SSD/SR/161/81.

5. The effect of stress waveform and hold-time on environmentally assisted fatigue crack propagation in a carbon-manganese structural steel. Report No. RD/L/N117/79.

6. A reappraisal of the in-service inspection schedule for the high pressure hydraulic components of Dinorwig Power Station. Report No. OED/STN/88/20024/S.

7. Investigation into the failure of Ffestiniog No. 4 turbine spiral casing, inlet half, Drawing N622/599 whilst on hydrostatic pressure test at Netharton Works. Report No. R/R/Mp. 12 by D W Hinchliffe.

8. A fracture mechanics analysis of the weld defects in the spiral casing to lower speed ring welds at Ffestiniog Power Station Units No. 1 and 2. Report No. NW/SSD/SR/14/77 by B L Bakie et all.

9. Dinorwig in-service inspection. Report No. NWR/SSD/85/0015/R.

The in situ repair of the bottom cover of a 300MW pump turbine at Dinorwig power station

V O MOSS CEng, FIMechE
First Hydro Company, Dinorwig Power Station, Llanberis, UK
G JONES CEng, MIMechE
Kvaerner Boving Limited, Doncaster, UK

1. ABSTRACT

Dinorwig Power Station commissioned in 1983 is among the largest pumped storage power stations in Europe. It was specifically designed to meet sudden increases in electricity demand and can generate up to 1320MW output within approximately 12 seconds.

This paper describes the following activities:-

i)   How a leak in a spiral casing cover joint of a large pump turbine deteriorated with time to such an extent that a major stripdown of the machine became necessary.

ii)   Methods adopted to try and address the problem i.e. with different types of sealing compounds and the methodology involved.

iii)   The development phase of the project, and how a full scale replica of the bottom casing of the Pump Turbine, was built which helped with the development and utilisation of new and existing insitu machining equipment, enabling the work to be carried out within a very limited space available.

iv)   In order to reduce the down time of a major stripdown the suction cone at the turbine was cut into pieces to allow for a bottom dismantling, instead of a more conventional approach which required upwards and sideways removal.

v)   The bottom cover had to be lowered with special lifting and handling equipment, access tools had to be designed to transport and handle the heavy components. A special machining device was installed to carry out the repair.

vi)   Planning, risk analysis, safety assessment and staff training were a major contribution to the success of the project.

2. INTRODUCTION

Dinorwig is one of two pumped storage plants operated by First Hydro an Edison Mission Company. The Dinorwig pump turbines and valves were designed by Kvaerner Boving Ltd. The station comprises of six vertical 300MW units connected to a single penstock. It is on one of these units that a leak developed in 1989 from the spiral casing lower cover joint. Attempts to arrest the leak in the spiral with injection methods were tried and failed. Eventually a stripdown of the machine had to be considered.

3. SPIRAL, STAYRING & COVERS DESIGN

The spiral casing is fabricated from carbon manganese steel plate and is welded to the cast carbon manganese stayring. In order to keep within transport limitations for access to the power station, it was necessary to split the spiral casing stayring assembly into quadrants. Jointing of the sections utilised cast flanges at the joints with pre-tensioned bolts. Shrink keys were also used at the stayring flange joints. Water tightness of the joint is achieved using toroidal acrylo-nitrile rubber rings. These are inserted into the water passage side of the spiral and stayring so that they can be replaced if necessary. The rubber rings are held in place using bolted on retaining plates. The critical point of such a sealing arrangement is where the radial joint of the spiral segments meets the circumferential rubber joint at the top and bottom covers. At these four points the radial and joint rings are butted against the circumferential rings.

The top and bottom covers are both one piece carbon manganese castings. To simplify the loading geometry, both connections with the stayring were made identical. Also, to avoid hoop loading on the stayring, the circumferential joint rings between the covers and the stayring are located as close to the water passage as possible and not under the connection flanges.

The bottom cover design necessitated its placement before the stayring and therefore cannot be removed after plant erection.

The suction cone fabrication is bolted to the underside of the bottom cover. Connection to the draft tube is by way of a slip joint. The cone is designed as a pressure containment capable of withstanding 60 bar, see Fig. 1.

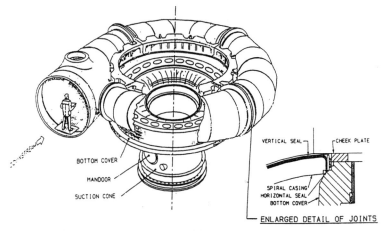

VERTICAL SEAL
CHEEK PLATE

SPIRAL CASING
HORIZONTAL SEAL
BOTTOM COVER

ENLARGED DETAIL OF JOINTS

BOTTOM COVER

MANDOOR

SUCTION CONE

DINORWIG POWER STATION
SPIRAL CASING & BOTTOM COVER

Fig. 1

4. ONSET OF LEAKAGE

The leak was detected on the spiral between 1989/90 and manifested itself as excess water flowing out of the plenum chamber drainage holes in the suction cone area. This deteriorated with time to such an extent that the drainage system from this area could not cope.

The leakage was measured by a Vee Notch weir collection system and the rate over the 4 year period was noted as follows:-

1990	- 270 litre/min
1991-92	- 900 litre/min
1993-94	- 1350 litre/min

The leakage was measured with the spiral at 60 bar pressure.

5. INVESTIGATION AND INTERIM SOLUTION

Over the 4 year period a series of holes were drilled into the spiral casing, and silicon rubber compound injected into the seal interface. This reduced the leakage for up to 9 months but the silicon rubber was eventually washed away with the high 60 bar pressure. A total of 15 holes were drilled into the spiral for the injection using different shore hardness of rubber compound. It was considered after the 3rd injection process that this was not good engineering practice and a longer term engineering solution be considered.

To try and pin point the exact location of the seal failure a series of ½" holes were drilled into the spiral plenum chamber and a TV camera plus video were inserted to locate the precise position (see Fig. 2).

With the system at full spiral pressure i.e. 60 bar, the leak was evident at a horizontal to vertical stayring to cover joint. From the TV pictures the seal housing appeared to have eroded away.

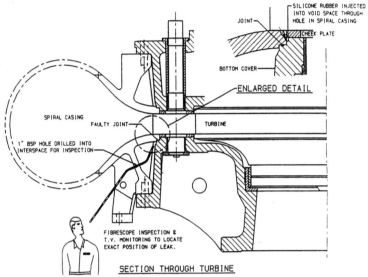

Fig. 2

6. ENGINEERING APPROACH

6.1 Options for Dismantling

The pump turbines are designed for upward and sideways dismantling at Dinorwig. Maintenance equipment is provided to allow the machines to be fully dismantled with the exception of the removal of the bottom cover. At the design stage, this was considered acceptable since the only items likely to require maintenance within the bottom cover were the cheekplates, guide vane bearings and wearing rings. These were all replaceable with the cover in place. The failure of the joint seal presented a challenge which had not been fully addressed in the station maintenance philosophy.

Options for dismantling were reviewed against two alternative repair scenarios.

- Minor Repair
  This would represent the situation where a localised repair could be carried out to the surfaces and the seals replaced.

- Major Repair
  This would be on the basis that one or both mating surfaces would require extensive on-site machining.

In the event of a minor repair it was considered that the turbine could be dismantled from the top using the maintenance equipment supplied. Combined with lowering the bottom cover by the maximum amount available, it was considered that access to carry out a limited repair and seal replacement was possible. The additional equipment needed for this option was minimal although the top cover dismantling would be extensive.

A major repair required more imagination and preparation. In this case, the top cover would not be disturbed but the bottom cover would be lowered to the floor of the suction cone area. Machining operations could then be carried out on the mating surfaces as required. The problems that needed to be addressed for this option were:-

- Suction cone prevented the cover from being lowered more than 230mm.
- Specialised machining equipment to be developed for working within the confined space.
- Handling of heavy equipment within the suction cone area i.e. typical weights bottom cover 63 tonnes, suction cone 8 tonnes.

The limited outage time allowed for the work meant that provision had to be made for the major repair scenario in order to prepare the necessary equipment and carry out supporting design activities.

7. ENGINEERING SOLUTIONS

7.1 Bottom Cover Lowering

It was determined that if the suction cone bottom 500mm was removed together with four projecting bosses, the remainder of the cone could be lowered into the draft tube and suitably supported. This would then allow the bottom cover to be fully lowered. A new section for the suction cone had to be designed in 4 pieces to be assembled and welded at site, see Fig. 3.

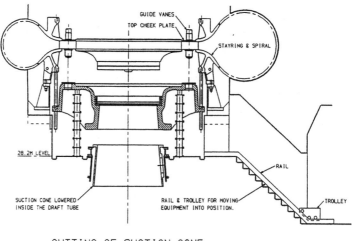

CUTTING OF SUCTION CONE

Fig. 3

A series of meetings were set up with the plant designers, site staff and contractors of the insitu machining equipment. From these discussions the most favourable solution appeared to be the proposal by Furmanite International Ltd.

Following a simulated demonstration at their UK works it was decided to consider the use of a circular self levelling milling machine (CSLM) developed by Furmanite International Limited for the on-site machining.

This machine includes a sophisticated control system that automatically adjusts the machining plane to align with any predetermined reference plane.

To ensure the work could be carried out safely at the plant a full mock up made of a fabricated steel structure was erected in the contractors workshops to demonstrate the suction cone and bottom cover. This allowed the insitu site machining equipment to be tested and the maximum man access could be studied for carrying out the work on site.

The original concepts had been proven in a power plant on the Snowy Mountain scheme in Australia but had to be modified to accommodate the Dinorwig situation. This machine gave an accuracy of 0.1mm of flatness over a 5 metre diameter of bottom cover.

7.2 Special Equipment and Handling

As part of the package of work, the sub-contractor developed handling equipment for the following tasks:-

-       Controlled lowering and raising of the bottom cover. Essential to avoid damage to the runner.

- Skid and trolly arrangement to bring equipment into the area through the access doorway and stairs.

7.3 Water Jet Cutting

To control the cut necessary on the suction cone, the use of a water jet cutter was adopted. As well as avoiding heat distortion of conventional cutting, this also allowed for a weld preparation to be generated during the cutting process.

7.4 Inspection

The first phase of the outage was to determine the extent of the repair. This was achieved by lowering the bottom cover within the available limit to allow inspection of the seal area. At this point, extensive erosion damage was evident and it became clear that a major repair was necessary.

Subsequent detailed inspection concluded that the original seal failure had occurred at a radial/circumferential joint intersection. The flow path, once established, caused the erosion to both the mating faces. There was little evidence of the circumferential seal remaining although it was noted that the four radial joint seals remained in good condition.

7.5 Repair and Re-Assembly

The procedure for the dismantling followed previously agreed method statements. The excellent degree of planning by all parties prior to the outage ensured the work commenced immediately and proceeded without serious problems.

The agreed repair to the seal surfaces was to machine the faces to achieve a sound metal condition. In the case of the bottom cover, it was decided to fit an

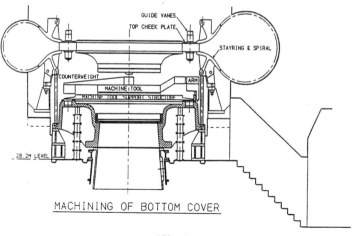

MACHINING OF BOTTOM COVER

Fig. 4

insert ring seal welded in place. The 'O' ring groove was then reinstated by site machining with due allowance made for the metal removed from the stayring face. Fig. 4 shows CSLM in position for machining bottom cover.

The short outage period required good project co-ordination to ensure parts and materials were delivered to meet the intensive 24 hour site activities.

Fig. 5 shows the bottom cover lowered with the machining equipment in operation.

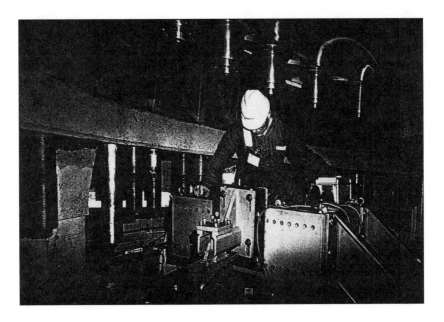

Fig. 5

One major area of concern was the re-building of the suction cone requiring extensive welding and stress relieving. The process was well controlled and distortion was found to be minimal.

Once the bottom cover had been fully raised, a pressure test was carried out to verify that a seal had been achieved. Due to the difficulty in attaining the necessary containment, this test was carried out at a pressure below normal working pressure. No problems were encountered.

The work was fully completed within the outage period, to budget and programme.

8. CONCLUSION

This paper demonstrates that teamwork and co-ordination between the designer and plant operator resulted in a well planned Engineering Solution to a major problem on an operational pumped storage plant in the UK. The work was completed during 1995 outage within 42 days compared with the predicted 90 days for a more conventional approach. This represented substantial revenue savings to the operating company.

Refurbishment and improvements in governing systems

KHOSA BSc, BE and **S BANTON** BEng
Kaerner Boving Limited, Doncaster, UK

Hydro Turbine governing systems have been rehabilitated and the performance improved by replacing the mechanical type or analogue electronic type governors with modern digital type electronic Governors and Electro hydraulic control. Improvements in the oil pressure systems includes the use of low noise positive displacement pumps, energy efficient motors, off-line filtration and in-house designed un-loader valves.

The optimum setting of governor parameters takes into consideration the influence of Derivative gain for fast response. Modern methods for Man Machine Communication have been developed.

1. INTRODUCTION

Governing of a prime mover (Hydro Turbines) was one of the first *closed-loop* system, developed over a century ago. The purpose of Governors is to control the output of the prime mover in such a way that its speed changes are controlled according to a pre-determined program and to regulate the power output in accordance with changes in grid frequency, with the help of Permanent Speed Droop, (**bp, ep**).

A number of hydro electric power projects have been in operation for the last **30** to **70** years, undergoing varying degrees of modernisation throughout their operational life. Many of these projects still have Hydro-generating sets controlled by mechanical governors or analogue type electro-hydraulic governors. These governors have many disadvantages namely:-

-       Non-availability of spares
-       Excessive leakage from the hydraulic control components (Distributing Valve etc.)
-       Constant internal leakage through the cast iron piston rings of the servomotors

- Increased Leakage from the Runner Distributing valve located in the runner hub.
- Some Governor parameters cannot be readily adjusted on Mechanical governors.
- Drifting of Transistorised circuit output of the Electro Hydraulic governors.
- Transistors sensitive to temperature in tropical locations.
- Non-availability of duplicate tripping.
- Dependence on Permanent Magnet Generator for speed measurement..
- Difficulty in interfacing with modern Distributed Control Systems and SCADA systems.
- Modes of operation, operating parameters and new features cannot be introduced in a cost effective manner, unlike the modern software based systems.

All the above mentioned disadvantages can be eliminated by full or partial rehabilitation of the Governing system. This rehabilitation can be executed independent of the rehabilitation and/or uprating of Turbines and generators.

One of the main aims of the investing in a new governor system is to enable the renovated machine to operate for another 25 to 30 years, with increased availability and flexible performance in accordance with the operating requirements. New governing can prolong the life of the turbine and generator by improving stability and by automatically avoiding unfavourable operating regimes.

2. OLD GOVERNING SYSTEMS

Examples of the old governing systems are briefly described as follows:-

2.1 Boving F-10 Mechanical governors

The F-10 Mechanical Governors were phased out in mid 1960's. Fig. No: 1a and 1b shows the typical arrangement of F-10 governor and its schematic, based on low oil pressure (**20** to **25** Bar) system, two stage governing concept and use of cast iron body for the regulating valve and guide vane servomotors. This design incorporates a mechanical feedback system, sometimes requiring extensive system of levers, shafts and linkages to suit the layout of the turbine. The result is often a control system with reduced sensitivity and an increased deadband.

Permanent speed droop, **bp** and temporary speed droop decay time constant, **Td** can be easily adjusted. However, the adjustment of temporary speed droop, **bt** requires shut-down of the unit. A limited range of standard sizes of regulating valve was available, typically 100mm or 150 mm. This limited choice often resulted in the valves being oversized for a given application, resulting in a further loss of sensitivity and a larger deadband.

2.2 English Electric Mechanical governors

The English electric mechanical governing systems were designed for single stage and two stage applications and an oil pressure range of **20-25** bar, to meet the requirements of servomotor capacity in terms of operating time and flows. These governing systems also required feed back, in the form of levers and linkages. Permanent Speed Droop, **bp** can be adjusted, while adjustment of other parameters (**bt** and **Td**) require shutdown of the unit.

The Mechanical Governor **Actuator Head** contains the following elements of the governing system:-

a.   Pendulum assembly and Pendulum Motor
b.   Dashpot (Temporary Speed Droop system)
c.   Permanent Speed Droop mechanism
d.   Speed Setting mechanism and motor
e.   Pilot Distributing Valve, Size: 2.5"
f.   Guide Vane Limit mechanism and motor.
g.   Guide Vane position and limit Indication system.

The Actuator Head, for 2-stage governing, as shown in Fig. No: 2, is mounted on a Hydraulic amplifier, which contains the pilot Servomotor and Main distributing Valve, of size: 10", and includes a facility for manual operation using a handwheel. Control Valves for emergency closing are mounted on the hydraulic amplifier. A floating lever, connecting the Pilot Servomotor, Main distributing valve and Main guide vane servomotors along with the return motion linkage completes the governing system. Control Block Diagram of Mechanical governor is shown in Fig. No: 4.

2.3 LMZ Electro Hydraulic Governor

The Electro Hydraulic Governors, manufactured up to mid 1970's by LMZ, Russia, was based on magnetic amplifier technology with limited additional features. The hydraulic part of the governor didn't have shutdown device, independent of the electrical/electronic circuits. This inadequacy was met by introducing an independent shutdown valve located between the governor and the servomotors. On later models, this type of governor was manufactured by Bharat Heavy Electricals Ltd., India. The author introduced a reliable shutdown directional control valve within the hydraulic system. The Governor electronics was modified based on Integrated Circuit technology with additional features such as frequency dead band, damping limits etc.

3. REHABILITATION & IMPROVEMENTS

3.1 Bhira Hydro Power Project, India

The 5 Units of Bhira Power Plant in India were commissioned in 1925-27, feeding the industries and domestic load of Bombay. An additional unit was later commissioned in the same station. The Pelton Turbines, Inlet Valves and Mechanical Governors were supplied by English Electric, UK. These machines were renovated in 1975-76 with new Runners, Inlet

Valves and Electro Hydraulic governors, retaining the shaft, bearings and the renovated generators. As a result, each unit was uprated by approximately 20%

The generation at this power plant is based on operation of all units for at least **8-10** hours every day at full load. The new Governors achieved the following :-

- Loading of each unit within **70** seconds as against **400-600** seconds.
- Synchronising within **10-15** seconds with the help of an auto synchroniser.
- Adaptation of the Governor for operation from a remote (Load despatch) station.
- Optimisation of Governor parameters.
- Reduced Speed rise and pressure rise, of the renovated unit for full load rejection, by reducing the closing time of deflectors to **1.25** seconds and increasing the closing time of Needles (jets) and improved sensitivity of the governor.

3.2 Jammu Tawi Canal Power Station, India

The **500** kW Generating sets at Jammu Tawi Canal power station, India is equipped with turbines and governors of East European manufacture. The performance of the governing systems had deteriorated over the years. No drawings were available to help rehabilitation. The aim was to introduce minimum changes in the governing system, due to the fact that the existing oil pressure system was operating satisfactorily. The mechanical governor was removed, retaining the servomotor. The following features were introduced :-

- Servo Valve and Electrical feed back on the existing servomotor
- Start/Stop valve
- Piping to suit
- Electronic Governor
- Toothed wheel and magnetic pick-up for speed measurement
- Interface with the existing control, for local operation from the governor cubicle and control room.

In fact, all the work of mounting the hydraulic control devices was carried out at site to suit the existing servomotor and shaft. The P&ID of the hydraulic control is shown in Fig. No: 3. The electrical feedback, mounted on the distributing valve, ensures the stability of the distributing valve opening in relation to the input signal to the electronic card, which drives the Servo Valve.

The two units were operated in isolated load with improved performance (1).. Frequency dead band was introduced for steady state operation, within a narrow band of +/- **0.3 Hz.**

3.3 Dual closing

A number of Kaplan Turbines, in Spain, were supplied during 1930-40's with a combined regulating ring and servomotor. This design didn't include servomotor cushioning at the end of the closing stroke. Cascade failure of breaking links was experienced on number of occasions. The breaking links were replaced by friction device of Ringfeder make. The cushion (slowing down the rate of guidevane closure), at the end of closing stroke was introduced in the regulating valve of the governor, to give improved performance and eliminate the failure of guide vane regulating system.

3.4 Ffestiniog, Wales

Ffestiniog Power station houses four units of **90 MW** Francis turbines, each coupled to a separate multistage pump via a clutch mechanism.

The existing governing system, for each unit, commissioned over 30 years ago, was manufactured by English Electric Co., UK as described in section 2.2.

First Hydro (owner of the power plant) decided to modernise the governors at Ffestiniog, but wished to retain the existing low oil pressure system, guide vane servomotors and oil piping. A cost effective solution was engineered, involving the following :-

a. Removal of the existing governor Actuator head, hydraulic return motion and various control devices on the hydraulic amplifier.

b. Fitting a new Governor Electro hydraulic Control unit , in a manifold block design, with its base matching the existing hydraulic amplifier.
    The new system consists of :

    Proportional valve (Moog)
    Start, Stop and Emergency stop valves
    Auto/Manual control valve
    Retard/Reset Control valve
    Timing throttles, Cartridge valves, Pressure Gauge, Isolating Valves etc.
    Fitting of an Feed back Transducer (LVDT) at back of Pilot Servomotor.

c. Replacement of cast iron piston seal of Pilot Servomotor with drop tight modern elastomer seal.

d. Fitting of feedback cam switch box in the Turbine pit for unit control requirements.

e. Digital electronic Governor, designed to suit the requirements of operation under generation, synchronous condenser and pump modes, supplied in 19" rack, fitted and installed in a   panel by First Hydro. The Electronic Governor has the following features :-

    -   Permanent Speed Droop - selection of two settings.
    -   Frequency Dead Band - selection of two settings.
    -   Normal and fast rate of loading.
    -   Guide vane limit control to suit various operating conditions.
    -   Maximum and Minimum Load Limits.
    -   Automatic changeover of governor parameters for smooth, disturbed and isolated operation.
    -   Automatic two limit settings, during start sequence, to optimise start-up time.

-       Parallel digital communications to the station control PLCs.

The function diagram of the refurbished governing system is as per Fig. No: 5 and Fig. No: 6 shows the Block Diagram of the governing system.

The rehabilitation of the governing system has resulted in the following improvements:-

a.   **Plant Availability**

Control of the guidevanes during start up is now under the action of the digital governor. Guidevane position is controlled precisely, to accelerate the machine as quickly as possible without causing undue mechanical or hydraulic stresses. Once rated   speed has been achieved the digital governor controls the machine speed in accordance with the grid frequency. The superior performance of the digital governor allows synchronising to be achieved consistently within the time dictated by commercial contracts. As a result the availability of the machines has been significantly increased and failure to synchronise incidence have been reduced.

b.   **Emergency Response, Spin Generate to Generate mode**

If an emergence low grid frequency is detected, a machine in the spin generate mode must be brought on line as soon as posible.
In order to achieve a faster transition from Spin Generate Mode to Generate mode, the governor begins opening the guidevanes before the Main Inlet Valve (MIV) is fully open. MIV position signals are fed into the governor to enable the governor to limit the guidevane opening to suit the MIV position. This feature reduces mechanical stress on the MIV components.
The resulting mode change considerably reduces the loading time of the machine . See Fig.No: 7.

c.   **Emergency Response to Low Grid Frequency from Part Load**

This facility enables the machine loading rate to be doubled when an emergency, low frequency condition is detected on the grid system. When an emergency low frequency condition occurs, a machine that is currently operating at part load immediately adopts full load operation (2). The change of power output is done at a fast loading rate of **12s** (3). See Fig. 8. This feature operates independent of the normal droop action.

d.   **Governor Controlled Normal Closing Rate**

During normal operation the closing of the guidevanes is controlled precisely by the governor to a rate which prevents the Spiral Casing Relief valve from opening. This feature dramatically increases the life of the relief valve components. A fast closing rate is still adopted for load rejection and tripping of the units.

e.   **Maximum and Minimum Power Limits**

The minimum power limit is particularly useful for the Ffestiniog machines as the rough running conditions experienced below 50 MW can be avoided.

f.   **Interface with Unit Control PLC - Remote Control**

The new governor includes a parallel digital communications link with the unit control PLC. This enables the governor to be controlled indirectly by a touch screen on the control desk and also remotely from the Dinorwig power station. See fig. 9.

g.   **Replacement Mechanical Components**

The replacement of cast iron piston seals with modern elastomer seals has resulted in negligible leakage in the Governor Electro Hydraulic Control Unit and Pilot Servomotor.
The components of the new control unit are all proprietary items and spares are readily available.

3.5 Dinorwig, Wales

Dinorwig Pumped Storage Power Station houses six Pump Turbines and Synchronous Generators.

The existing governing system consisted of:-

a.   Analogue electronic governor of ABB make.
b.   Electro hydraulic control unit type E-40 of KMW/Boving make.
c.   A return motion linkage between the guide vane servomotor and the electro hydraulic control unit is provided to complete the closed loop in the second stage.

First Hydro decided to replace the governor, retaining the existing oil pressure system, guide vane servomotors and oil piping. The need for replacement was dictated by commercial requirements and renewed life of over 20 years.

The cost effective modernisation program was adopted as follows:-

a.   Removal of E40 SM3 Actuator and replacement by E40 SM4 actuator of Kvaerner make.
b.   Removal of analogue turbine and pump governors from the governor cubicles.
c.   Mounting new ABB make digital governor within the existing cubicle.
d.   Replacement of position transmitters.
e.   New man-machine interface in the control room.

The most significant feature of the E40 SM4 actuator is a servo valve system with electrical feedback resulting in precise positioning of the actuator. The feedback device is a 150mm stroke LVDT fitted to the actuator.

The ABB digital governor, designed to control all modes of operation, is supplied in rack mounted configuration for easy assembly within the existing governor cubicle. The governor incorporates the following features:-

a.   Permanent droop - selection of 4 settings
b.   Frequency dead band - pre-set adjustment
c.   Power set point
d.   Asymmetric droop about the frequency set point
e.   Frequency set point
f.   Minimum and maximum power limit
g.   Banned region of generator operation with hysteresis function for satisfactory operation during loading/unloading sequence
h.   Automatic changeover of governor parameters for smooth, disturbed and isolated operation. (Currently not used for commercial reasons)
i.   Part load emergency response - similar to that described for Ffestiniog
j.   Spin Generate to Generate mode change, with a rapid loading rate.
k.   Manual control independent of governor software and servo valve
l.   Test facility that provides frequency injection and power supply monitoring.
m.   Pump control in accordance with the Net Head/Frequency/Guidevane Position curves.

The existing man-machine interface (MMI) on the control desk consisted of analogue meters and push buttons. The new MMI is in the form of a programmable video touch screen panel. This panel is connected to the governor via a serial line to allow operation of a hydro-set from the control desk.
The following controls and indications are available, at the control desk:-

a.   Banned region selection
b.   Power set point and indication
c.   Frequency set point and indication
d.   Maximum power limit set point and indication
e.   Minimum power limit set point and indication
f.   Guide vane position indication
g.   Guide vane limit position set point and indication
h.   Auto/manual control and indication
i.   Permanent speed droop setting and indication
j.   Frequency dead band in-out control and indication
l.   Part load response selection, with a rapid loading rate.
m.   Governor fault alarm.

Separate push buttons and indication have been provided for operation under "manual" control, independent of the video control panel. The existing actuator pilot servomotor position indication (analogue) type has been retained.

The client ordered one set of spares for the Electronic governor and one extra governor to ensure availability of the governor, for satisfactory operation, for the next **20** years

4.   OPTIMUM SETTING OF GOVERNOR PARAMETERS

The performance of a governed hydro-set, employing the Mechanical or Analogue Electronic governors, is limited due to the Temporary Speed Droop parameter. The response of the governed hydro-set can be improved, by the influence of the Derivative gain of the three term PID control, introduced in the modern governors (4). The improvement can be best explained by the new settings introduced at Ffestiniog Power station.

Governor parameters of the Old Mechanical governor:-

Temporary Speed Droop, **bt**	**0.6** p.u
Dashpot Time Constant, **Td**	**11** secs.

Governor parameters of the New governing system:-

Proportional Gain	**3.5**
Integral Gain	**0.4** 1/sec.
Derivative Gain	**3.8** sec

The Derivative gain can be increased to **0.5 Ta**, where **Ta** is the Machine Time Constant. A higher setting of Derivative gain allows further optimisation of Proportional and Integral gains. As a guideline, the response of the Temporary Speed Droop type governor is proportional to the product of **bt*Td,** and that of PID type governor is inversely proportional to the **Integral Gain**.

5   OIL PRESSURE SYSTEMS

The rehabilitation of the Oil pressure system is based on the following aspects:-

a.   Retain the existing oil pressure system. Replace the defective components. Introduce off-line filtration to improve the cleanliness of the oil. Replace the Piston rings of the servomotors with the modern elastomer type seals, if required.

b.   Total replacement of the Oil pressure system with High pressure (say **100** to **160** bar normal pressure), incorporating Piston accumulator and Nitrogen bottles. This option does not require compressed air system and uses less quantity of oil due to the reduced size of the servomotors.

c.   The most common option, for Kaplan Turbines, is to retain the existing low pressure system for Runner Blade control and provide high pressure system for Guide Vane control.

d.   Improved performance is achieved by the selection of low noise positive displacement pumps of the internal gear type, energy efficient electric motors, pressure compensated variable displacement pumps which do not require un-loader valves, off-line filtration, low level and high temperature switches, located in the oil sump tank, to trip motors. The noise level achieved is within **75 dBA.**

6. GOVERNOR ELECTRO HYDRAULIC CONTROL

The modern governor electro hydraulic control is generally based on the selection of Fluid Power devices and components, meeting the requirement of CETOP. This method is cost effective and the spares are easily available. The hydraulic governing systems supplied by the Kvaerner Group are designed to suit individual station requirements and the main devices for example the Proportional valves are sized for optimum performance.

6.1 Replacement of F-10 Mechanical Governor

The F-10 Governor assembly, as shown in Fig. No: 1a, is removed and all control linkages from this governor to the Regulating Valve are removed. The control chamber of the regulating valve is operated by a Proportional Valve, with an electrical feedback (LVDT) provided on the regulating valve. The concept is similar to that shown in Fig. No: 3.

6.2 Governing systems for Kaplan and Pelton Turbines

Many of the traditional mechanical items of a hydraulic control system are now included in the governor software. For example, mechanical cam For Kaplan Turbines, for runner blade to guide vane relationship, is replaced by an electronic cam within the software of the governor. This 'cam' takes the form of a look-up table within the software, containing data derived from the model test. Similarly the mechanical cam for needle to deflector relationship, in Pelton turbines is replaced by electronic cam.

In both the above examples close loop position control systems are now used, utilising proportional electro-hydraulics and electrical position feedback.

7. ACKNOWLEDGEMENT

The authors wish to express their gratitude for the help and co-operation which they received from the engineers of First hydro and thank their employer, Kvaerner Boving Limited, for the permission to publish this paper.

8. REFERENCES

1. **Khosa, J. L** - " Rehabilitation of Jammu Canal Power Station Governors ", BHEL Report.
2. **Aris, F. C, Jones, G. R and Price, G.**- " Exploiting New governor Control At Ffestiniog Power Station ",Water Power & Dam Construction,Uprating and Refurbishing Hydro Power Plants V Conference, Nice 1995.
3. **Banton, S.**- " Commissioning Report Of Dinorwig governors ", Kvaerner Boving Document.
4. **Khosa, J. L.**- " Influence of Derivative Gain on the optimisation of Proportional and Integral Gains ", Kvaerner Boving Document.

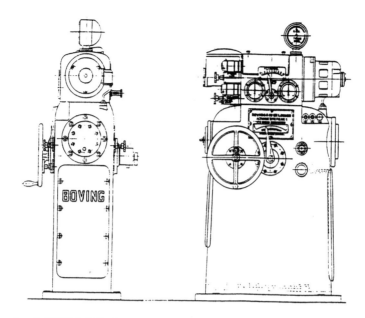

Fig. No: 1a- BOVING MECHANICAL GOVERNOR TYPE F-10

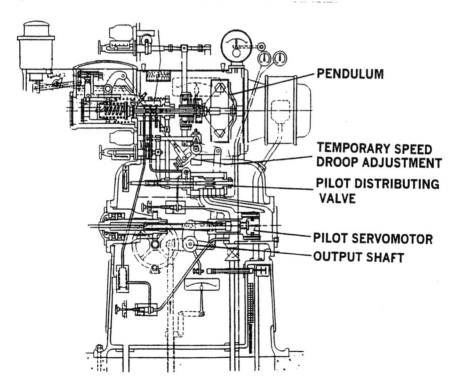

Fig. No: 1b- SCHEMATIC ARRANGEMENT OF F-10 MECHANICAL GOVERNOR

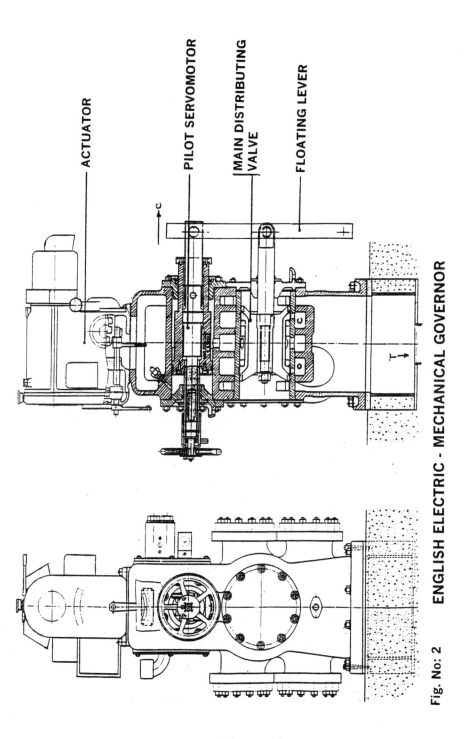

ACTUATOR

PILOT SERVOMOTOR

MAIN DISTRIBUTING VALVE

FLOATING LEVER

Fig. No: 2   ENGLISH ELECTRIC - MECHANICAL GOVERNOR

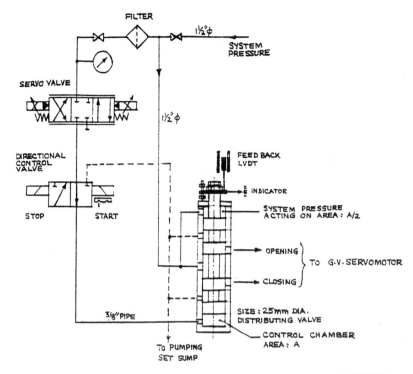

Fig. No: 3 - JAMMU TAWI CANAL REFURBISHED GOVERNOR ELECTRO
HYDRAULIC CONTROL - P&ID

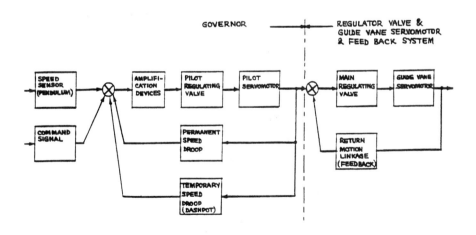

Fig. No: 4 - CONTROL BLOCK DIAGRAM OF MECHANICAL GOVERNOR

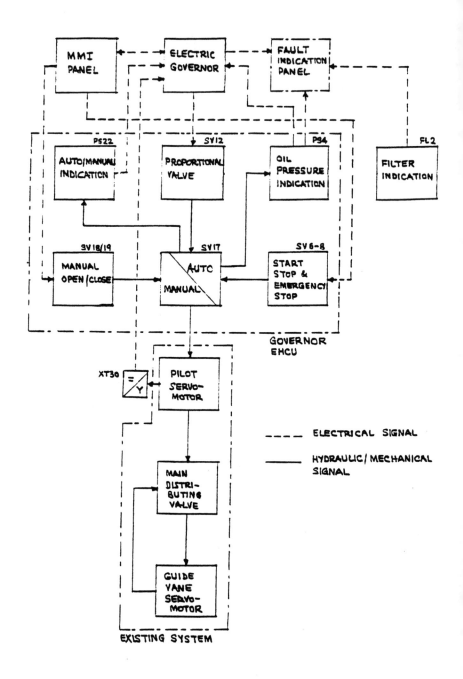

Fig. No: 5 - FUNCTION DIAGRAM OF REFURBISHED GOVERNOR

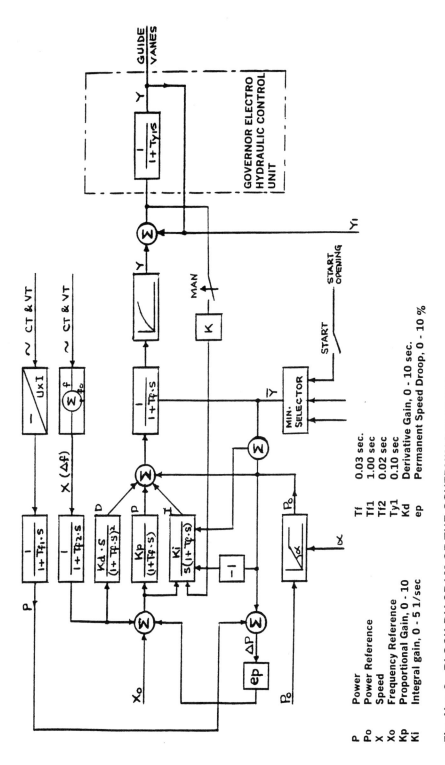

Fig. No: 6 - BLOCK DIAGRAM OF THE GOVERNING SYSTEM

P   Power
Po   Power Reference
X   Speed
Xo   Frequency Reference
Kp   Proportional Gain, 0 - 10
Ki   Integral gain, 0 - 5 1/sec

Tf   0.03 sec.
Tf1 1.00 sec
Tf2 0.02 sec
Ty1 0.10 sec
Kd   Derivative Gain, 0 - 10 sec.
ep   Permanent Speed Droop, 0 - 10 %

Fig. No: 7

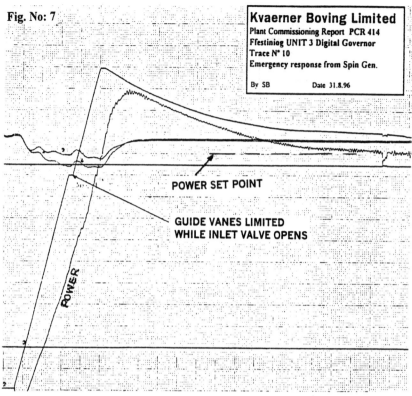

Kvaerner Boving Limited
Plant Commissioning Report PCR 414
Ffestiniog UNIT 3 Digital Governor
Trace N° 10
Emergency response from Spin Gen.

By SB   Date 31.8.96

POWER SET POINT

GUIDE VANES LIMITED
WHILE INLET VALVE OPENS

POWER

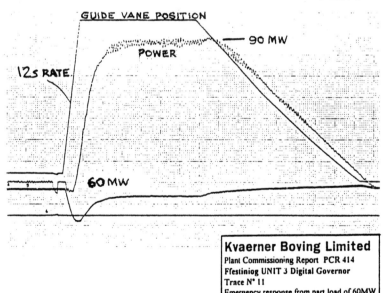

GUIDE VANE POSITION

90 MW

POWER

12s RATE

60 MW

Fig. No: 8

Kvaerner Boving Limited
Plant Commissioning Report PCR 414
Ffestiniog UNIT 3 Digital Governor
Trace N° 11
Emergency response from part load of 60MW

By SB   Date 31.8.96

FIG. 9 - Communications System and Remote Control - Ffestiniog Power Station

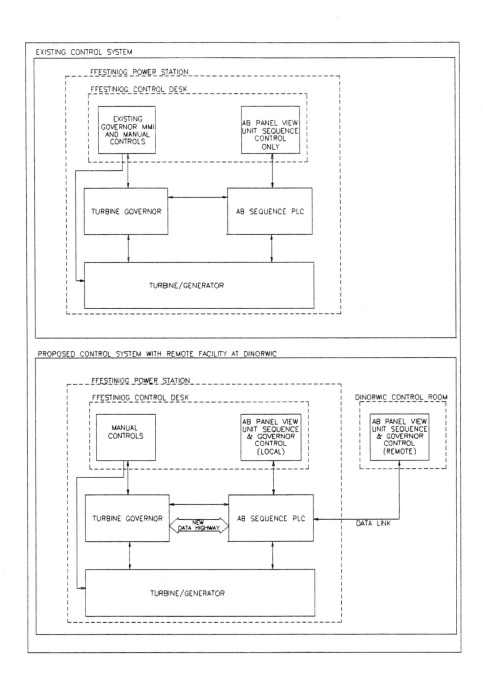

Severe hydro-mechanical vibrations following the upgrading of a 30MW UK hydro station – a case study

J GIBBERD MEng, CEng, MIMechE
WS Atkins Science and Technology Limited, Bristol, UK

Following upgrading of the 30MW Maentwrog hydro plant, severe vibration occurred in the penstocks near full load. This was found to be due to a hydraulic resonance driven by a draft tube vortex in the new Francis sets. The vibration posed a fatigue threat to pipeline integrity, and a 20MW load limit was imposed. This paper discusses the unusual combination of factors which led to the problem, and the design of a package of measures to enable a safe return to full load running. These measures included the provision of an air injection system to reduce the pressure oscillations across the runner, and the installation of a modified pipe support arrangement to reduce the pipe wall stresses.

1. INTRODUCTION

Maentwrog power station is located in North Wales, UK, within the boundaries of the Snowdonia National Park, and is owned and operated by Magnox Electric plc. It was constructed in 1928, and underwent a major upgrade and refurbishment programme starting in 1991, which included the construction of a new dam, refurbishment of the pipelines, and the installation of uprated turbogenerators, new controls and switchgear. A brief description of the plant is given in Section 2 below.

As part of the upgrade, two new 15MW horizontal axis Francis type reaction turbines were installed, replacing the original 4 x 6MW twin-jet Pelton impulse turbines. During commissioning, power oscillations of +/-2MW were observed on attempting to run both units at full load, accompanied by significant cavitation noise and mechanical vibration at the powerhouse, and it was not possible to run the units at greater than 2 x 14.5MW. Away from the powerhouse, vibrations were also observed in the pipelines, those in the uppermost lower head region being of particularly high amplitude.

Above the support saddles, the vibration amplitude was of the order of several millimeters, and this gave rise to concern for the fatigue life of the pipes. The maximum permissible output of each unit was therefore restricted to 10MW for safety reasons, at which load the vibration amplitude was considerably lower. The station output was thus reduced by 10MW, or one third of its design output.

Early in March 1994 a programme of test and analysis work was undertaken, with the aim of understanding the vibrational behaviour and its implications for fatigue life, diagnosing the source of the excitation and reviewing potential solutions. This paper discusses the unusual combination of factors which led to the problem, and the design and implementation of a package of measures to enable a safe return to full load running.

2. DESCRIPTION OF THE PLANT

The water comes from Lake Trawsfynydd via a 2.8km low-pressure (LP) pipeline which emerges at the top of a steep hill immediately behind the power station. The LP tunnel terminates in a valve house which marks the start of the high-pressure (HP) penstock system. At this point the line bifurcates into a 1.87m diameter pipeline and a 1.52m pipeline, both of which are fitted with butterfly valves. The two HP penstocks run approximately 450m from the valve house down to the station.

Fig 1 - View of pipes at the area of worst vibration

The section of interest is that between the HP valve house and the first anchor block downstream, in which the most severe vibrations were observed. The general layout is shown in the photograph in Fig.1, looking downstream from the valve house with the 1.87m pipe on the left. The HP pipelines are made from rolled steel plates, gas welded longitudinally and rivetted together at overlapping joints. They are supported on concrete saddles at 15.2m intervals via lubricated steel saddle plates. The wall thickness is 11.1mm for the 1.52m pipe and 12.7mm for the 1.87m pipe in this section, but the thickness increases in steps along the HP line to reach 26mm for both penstocks just outside the station.

The power station is equipped with two 15MW horizontal-axis Francis type reaction machines, operating under a maximum gross head of approximately 190m with a nominal full-load discharge of 9.25m³/s. The layout of Unit 1 is shown in Fig.2. The units run at 600 rev/min, and are synchronised onto a strong grid. They are not equipped with governors, load control being achieved by manual adjustment of the guide vane position. The machine centrelines are set at the 7.18m AOD level, with the mean tailrace being at approximately 4.3m AOD. At this height the machines were susceptible to cavitation, and orifice plates were fitted into the draft tubes in order to raise the back pressure.

Fig 2 - View of the Unit 1 turbine

3. DIAGNOSIS OF PROBLEM

3.1   Site observations

With both units at 15MW, harmonic pressure fluctuations of approximately +/- 1.5 bar were observed in the penstocks at approximately 3.0Hz, which corresponded in phase and frequency to loud cavitation in the draft tube and power surges on the MW meter. Strong periodic cavitation was audible in the draft tube just under the runner, and the pipelines near the valve house were undergoing a forced ovalling vibration at the same dominant frequency.

The only major change to the pipelines during re-planting was the change in the downstream hydraulic boundary condition from discharging freely to atmosphere (for the Pelton wheels) to being coupled to reaction turbines. Prior to the re-planting there had been no reported vibration. The clear periodicity and frequency correlation between the valve house vibration and the turbine pressure oscillations and draft tube cavitation suggested the resonance of a natural hydraulic mode of the penstock system, driven by a hydraulic instability in the draft tube. Such problems with the resonance of pressurised piping systems have been well documented in the past, following the Lac-Blanc / Lac Noir incident, and later the Bersimis case (1), their usual manifestation being 'panting' or 'singing' penstocks. The existence of a hydraulic resonance was supported by reports from the turbine manufacturer that pressure surges at around the observed frequency on site had been experienced during model testing.

3.2   Tests and analytical modelling

A series of hydraulic tests was carried out in order to confirm the diagnosis. Dynamic pressure was measured at four locations along the pipeline from the turbine main inlet valve up to the valve house, over a range of loads. The pressure spectra with load in the 1.87m diameter pipe are shown in Fig.3. At loads less than 2 x 15MW, the pressure forcing function was broad-banded, however at 2 x 15MW a clearly defined peak of large amplitude arose at around 3.0Hz. This correlates with the commissioning report that the units could not be run at 2 x 15MW. The variation of rms vibration acceleration with load of the 1.87m pipe near the valve house is shown in Fig.4; it can be seen that the resonance doubled the vibration amplitude.

The existence of a hydraulic resonance was confirmed by analytical computer modelling, using the impedance method (2) to determine the natural frequencies of the pipeline system. The modelling identified a mode close to the measured frequency of 3.0Hz. The predicted mode was the seventh mode between the station and the surge tower, which correlated well with the measured mode as shown in Fig.5.

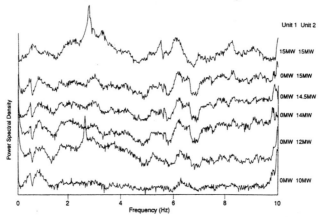

Fig 3 - Pressure spectra with load near the valve house

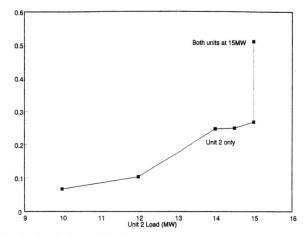

Fig 4 - Variation of RMS acceleration of 1.87m pipe with load

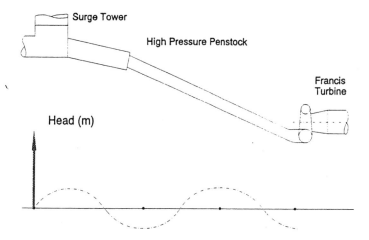

Fig 5 - Dominant hydraulic mode of penstock system as predicted and measured

3.3 Pipeline response: stress and fatigue analysis

Having established the cause of the vibrations, it was necessary to investigate the pipe dynamic response and carry out a stress and fatigue analysis.

The pipes were observed to oval considerably under their self-weight (~ 50te per span) at the support saddles, and superimposed on this static deflection the pressure oscillations forced a dynamic ovalling response. The interesting point at Maentwrog was the great flexibility of the pipes in the ovalling mode, due to the low wall thickness in relation to diameter. Dial gauge tests demonstrated that a change in internal head of only 1m

(0.1bar) gave rise to a radial displacement of almost 1mm.

The region of most concern for fatigue was the horn of the saddles, where the high load, thin walls and low arc of support (90°) combined to produce high circumferential bending stresses, both static and dynamic. At all other points, the vibrations could be tolerated since negligible dynamic stresses were developed. Calculations using the methods of BS 5500 for pressure vessel support saddles (3) indicated peak static stresses approaching twice the yield stress in this region, and this was later confirmed by finite element modelling. The original saddle design can be seen in Fig.6, which also shows the bulging at the horn predicted from the finite element model. Note that in practice, these peak membrane bending stresses are highly localised at the horns and will have been redistributed by local plasticity.

The vibrational mode shapes of the pipelines were also measured during the hydraulic tests, from which a correlation was derived between vibration level and dynamic stress. A preliminary fatigue assessment predicted failure within a year or so at 2 x 15MW, with a maximum rms stress level of approximately 12MPa.

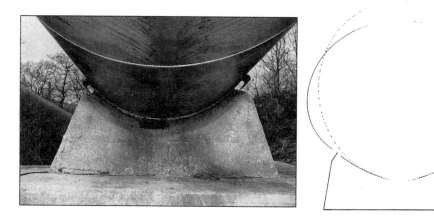

Fig 6 - Original saddle design, with predicted deflected shape from FE modelling

4. DISCUSSION OF THE CONTRIBUTORY FACTORS

The interesting aspect of the Maentwrog case is the way in which a large number of factors, none of which are unique to this plant, combined to cause an unusually severe problem. Considering these factors in more detail:

4.1 Hydraulic resonance: turbine-pipe interaction

The vibration problem was initiated by the replacement of the impulse wheels with reaction machines. This not only altered the downstream hydraulic boundary condition on the penstock system (and hence the hydraulic natural frequencies and mode shapes), but also provided an oscillating excitation mechanism in the form of the hydraulic instability in the draft tube.

Many Francis turbines exhibit harmonic pressure oscillation (surging) in the draft tube due to the well-known rope vortex. The strongest excitation usually arises at part load (40% to 70%), however at Maentwrog the worst vibrations surprisingly arose at or near full load. This was because the vortex frequency is a function of load, typically as shown in Fig.7; the resonance with the pipeline system did not therefore occur until high loads when the vortex frequency happened to lock onto a natural pipeline mode. It is nonetheless unusual to encounter such a strong resonant response at full load.

The question arises of whether the resonance could have been predicted prior to upgrading. It would have been possible to compare the surge frequencies from turbine model tests with a calculated frequency response of the penstock system, however even if a potential resonance had been identified, it is notoriously difficult to predict the magnitude of the resulting pressure oscillations. In any case, the pressure oscillations at Maentwrog were of small magnitude due to the great flexibility of the pipes, and would probably not have given rise to concern.

The vibration problem at Maentwrog was partially exacerbated by the runner setting above the tailrace and the increased cavitation. Ironically, this is one parameter which could have been controlled at Maentwrog, since it was necessary to excavate and construct completely new foundations to accommodate the Francis sets. For most upgrades, of course, it is not cost-effective to alter the turbine casing elevation.

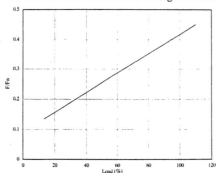

Fig 7 - Typical variation of draft tube vortex frequency as function of load for a Francis turbine

4.2   Design of penstocks & supports

The original design of the Maentwrog pipes and supports also contributed to the fatigue problem in the following manner:

(i)   In common with many others, the pipes were designed for a nominally constant working stress against the internal pressure head. This approach paid insufficient attention to stiffness in the upper low-head regions, where the pipes become thin-walled and susceptible to vibration.

(ii)   Design codes for steel pipelines have always tended to treat the pipes as simple beams, and the wall thickness is checked for longitudinal bending stresses at the supports. For larger diameter pipes which have a greater bending modulus, this approach has two effects: it leads to smaller thickness to diameter ratios, which renders them more flexible, and it allows longer spans, which causes greater loads at the supports. Modern pressure vessel support codes (e.g. BS 5500) suggests that this is an inadequate approach, since the bending modulus is reduced over a pipe support where a more complex stress field exists.

(iii)   Another common feature of early pipeline designs is the use of a 90° arc of support for the pipes. As outlined above, this can lead to high circumferential bending stresses at the horns. Again, modern practice is different and would indicate the use of a 120° minimum arc, or occasionally 150°.

The traditional design approach is perfectly adequate provided the loading is static, in which case any local areas of high stress can generally be relieved by plasticity without risk of pipeline rupture. When dynamic loading is present, however, a number of factors can combine to promote fatigue. Vibration would not have been envisaged at the time of designing the Maentwrog pipes, and indeed even modern codes give negligible guidance for designing against fatigue. The following all contributed to the fatigue problem at Maentwrog:

- the large number of cycles which can occur under vibrational loading (typically $>$ 10^9), which leads to fatigue endurance stresses of the order of only a few MPa rms;
- the high mean stresses arising from weaknesses in the traditional design approach, which reduced the fatigue endurance;
- the presence of welds with inherent defects, in regions of high stress;
- the potential for fretting, corrosion and pitting from weathering;
- uncertainties in the condition and material properties of the pipes;
- the presence of stress concentrators, particularly at welds and around rivetted joints.

5. SOLUTIONS

5.1 Air injection

The first approach was to attempt to reduce the excitation mechanism. The two common methods for dealing with draft tube surging are draft tube inserts and air injection. A comprehensive review of various systems reported in the literature is given in (4) along with a discussion of their effectiveness and influence on operation and efficiency. On the basis of this review, air injection was considered a more cost-effective method since it could more readily be optimised at site without the need for model tests. In addition, it did not involve the installation of structures in the draft tube, which could themselves have been susceptible to vibrations, particular in the high cavitation environment. (There are a number of reported cases where inserts have fatigued rapidly and been washed away).

Two redundant tappings were available on the turbine head cover, originally intended to allow for vacuum admission. A test rig was installed on Unit 2, capable of supplying a maximum of 28l/s of compressed air at a pressure of 1.5bar. Fig.8 shows the significant reduction in draft tube pressure oscillation achieved at 2 x 15MW. Air injection significantly reduced the cavitation noise at the powerhouse, at all loads.

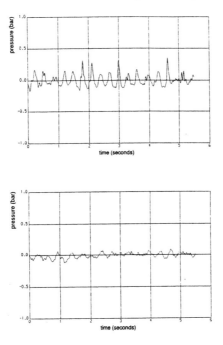

Fig 8 - Effect of air injection on draft tube pressure surging at 2 x 15MW with 28 l/s air injection

Fig.9 shows the rms vibration on the 1.87m pipe as a function of load and air injection flowrate. It can be seen that the initial 8l/s causes a significant reduction, but that thereafter the benefit becomes minimal. Using the previously derived correlation between vibration and stress, it was determined that a reduction in rms acceleration of >3.0 was required in order to remove the fatigue threat. It can be seen from Fig.9 that a reduction of approximately 2.0 was achieved, and pipeline structural modifications were therefore necessary in order to achieve a safe return to full load running.

A permanent air injection system is now installed on both units, with a capacity of 22l/s. The air supply is turned on by limit switches operating off the guide vane linkages. To ensure the units are never run without air injection, the turbines are shut down automatically in the event of loss of supply.

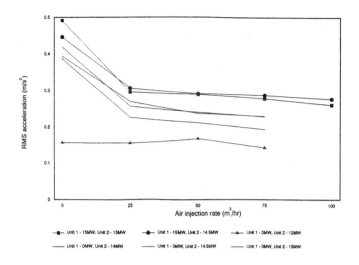

Fig 9 - Effect of air injection on pipeline vibration levels

5.2 Modified support arrangement

5.2.1 Feasibility study

Following the air injection tests it was clear that it would also be necessary to address the issue of high static and dynamic stresses at the saddles horns. A feasibility study considered the options of local pipe stiffening, the installation of new intermediate saddles, and the total replacement of the pipe. The test and analysis work indicated that it was only necessary to modify the 100m length between the valve house and the first downstream anchor block. Further downstream, the pipe walls become thicker as the next pressure antinode is approached.

> Local stiffening by means of rings local to the saddles was rejected due to the difficulty of providing sufficient additional stiffness close enough to the saddles to reduce the stresses. In addition, it was desired not to weld to the pipes given their age and uncertain manufacturing procedures, and there would have been difficulties in providing a bolted clamp joint of adequate capacity. The cost of removing the old pipes from the site (which is remote with difficult access) and installing totally new pipes over the 100m length of interest would have been ~£750,000; approximately three times that of installing new intermediate saddles.

5.2.2 Design of new support saddles

The new saddle design is shown in Fig 10. Twelve new saddles were installed (6 on each pipe), comprising steel fabrications bolted together and jacked into place on new reinforced concrete foundations. The new saddles have a 150° arc of support, and were installed to take 75% of the span load, thereby relieving the fatigue problem from the original saddles.

This option avoided the need to demolish the old saddles.

Fig 10 - New support saddle design

The new saddles were designed to enable the pipe walls to achieve a minimum fatigue life of 60 years (2.0×10^9 cycles) in conjunction with air injection. This was commensurate with the required life of the other station plant. A number of interesting problems had to be overcome in the design:

Design for high-cycle fatigue of 70-year old welds

A fatigue analysis was necessary to substantiate the design of the new saddles. This had to address a number of issues:

- uncertainties in the analysis of dynamic stress
- the high number of cycles
- the lack of high-cycle fatigue data
- the presence of welds of unknown quality
- the effects of fretting, pitting and corrosion
- the effect of the vibration which occurred following commissioning, but prior to the setting of a load limit

The stress field in the saddles regions is complex, and the original correlation between rms vibration and dynamic stress was relatively crude. Since the number of cycles is so high, the fatigue endurance stress is very low (of the order of a few MPa) and a highly accurate stress model was therefore required. In the absence of suitable guidance from design codes, software developed by the University of Strathclyde for the design of pressure vessel supports was employed (the theory of which is outlined in (5)).

Conventional steelwork design codes give little guidance much above 10^7 cycles, and it was therefore necessary to adopt a high-cycle fatigue methodology from the nuclear industry (7). This procedure is a conservative assessment methodology, supported by test data, which extends the fatigue S-N curve up to 10^{12} cycles. It is interesting to note that contrary to the normal assumption for structural steels, there is no clear fatigue limit and the endurance stress continues to fall with increasing cycles. The mean minus three standard deviation design curve for Class F butt welds in rolled plates is shown in Fig. 11.

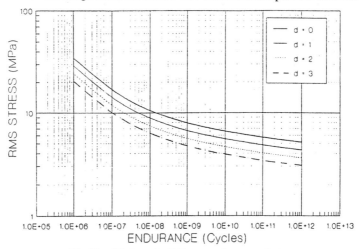

Fig 11 - High cycle fatigue endurance data

A sensitivity study identified that the results are sensitive to the wall thickness, and a programme of ultrasonic thickness measurements was therefore undertaken to establish the actual values. This programme also aimed to establish the general quality of the original 1928 gas welds, and the nature, size and extent of any inherent flaws, in support of the fatigue design of the new saddles.

Design for out-of-roundness

The original pipes were not perfectly round, due to their construction method which consisted of gas welding rolled steel plates together longitudinally (three plates were used for the 1.87m pipe, each one thereby forming a 120° arc, and two plates for the 1.52m pipe). Jacking trials were carried out, and due to the flexibility of the pipes it was found that they tended to conform quite readily to the shape of the saddle, even at relatively low loads. On some saddles, however, it was necessary to locally grind the horn region and to fit packers between the two halves of the support saddles in order to achieve a reasonable fit. This was a trial-and-error process carried out on site. The distortion of the pipe in conforming to the saddle shape induced static stresses, however the resulting good fit eliminated the dynamic component under pressure oscillations.

5.2.3 Installation of new saddles

The concrete bases were first excavated and the re-bar cage cast in, including a lower steel baseplate set at the slope of the pipe. The saddle halves (each weighing approximately 360kg) were then aligned and bolted into position. A number of interesting problems had to be overcome during the installation:

Site access

The access road to the site is a steep and rough track, and could not be metalled for environmental reasons because the site is within the National Park boundaries. There was also no vehicular access down between the pipes, and it was necessary to install a mobile crane at the valve house to lift the saddles over the pipes, and a tracked dumper to transport the saddles down between them.

Achieving a uniform load between the new saddles

In order to meet the required fatigue life, each new saddle was required to carry 75% of the load with a +25% / -15% tolerance. In order to achieve this load distribution, flat 'pancake' type jacks were inserted under the saddle baseplates and connected to a common hydraulic circuit. All the saddles on each pipe were then jacked up simultaneously to the required load. The saddle baseplates were monitored with dial gauges to ensure that no settlement occurred whilst the studs were tightened up, the jacks removed and the baseplates grouted in.

6. CONCLUSIONS

A number of lessons can be drawn from the Maentwrog experience, which may be of use for future upgrades:

- There is always the potential for hydraulic resonance, even in systems where there may not previously have been a problem. This is important when planning any significant changes to the hydraulic characteristics of an existing pipeline system. At Maentwrog, the replacement of impulse wheels with reaction sets altered the downstream hydraulic boundary condition, and introduced a new oscillating excitation mechanism. Seemingly less dramatic changes, such as installing new runners into existing casings, could also alter the hydraulic characteristics.

- Hydraulic instability (e.g. draft tube surging) is not necessarily only a part-load phenomenon, and resonance can occur at or near high load due to lock-on of the vortex frequency with a pipeline system natural mode. Where the cavitation margin is an influence, one must also be aware of the effect of other parameters, for example changes in head, temperature, and water chemistry, all of which can affect the onset of instability. At Maentwrog these factors gave rise to non-repeatability in the onset of the maximum vibration on separate tests under nominally identical operating conditions. In order to avoid running in the region of worst cavitation, it has been necessary to provide a safety margin and limit the maximum load to 15MW - 14MW on unit 1 and 2 respectively.

- Air injection can significantly reduce draft tube surging. A reduction of 50% in rms pressure levels was achieved at Maentwrog, with a flowrate of 0.2% of the gross water flow. If the possibility of hydraulic instability or resonance has been identified, it is worth making provision for head cover air injection points at the design stage. At Maentwrog this proved to be a straightforward solution.

- Older pipeline designs may have paid insufficient attention to stiffness in the upper low head regions, and can be susceptible to vibration from changes in internal pressure. The Maentwrog pipes proved to be highly responsive, with a change in internal head of only 1m leading to a radial displacement of approximately 1mm. This makes it difficult to remove the vibration, since some degree of pressure fluctuation is inevitable with reaction machines. Pipeline structural modifications may then be necessary.

- Older designs may also provide an arc of support of only 90°, which in conjunction with high spans and low thickness to diameter ratios can induce high circumferential bending stresses at the horns (approaching twice yield at Maentwrog). This is not a concern in the absence of vibrations and can be tolerated. If the pipe is subject to vibration, however, the dynamic stresses can pose a high-cycle fatigue threat.

- Pipeline design codes give little guidance on analysis or design against fatigue loading, and conventional structural codes give little fatigue data much above 10^7 cycles. In order to solve the Maentwrog problem, design tools have therefore been adapted from other industries. Finite element software was adapted from the pressure vessel industry to analyse pipe wall stresses, and a high-cycle fatigue analysis methodology has been transferred from the nuclear industry.

- The Maentwrog experience has demonstrated the feasibility of installing new saddles to relieve highly stressed pipe supports, and support the pipe in a more benign manner.

ACKNOWLEDGEMENTS

The author wishes to acknowledge Magnox Electric plc for their permission to publish this paper, and all the station staff at Maentwrog for their assistance during the site tests, construction and commissioning. The author also acknowledges the previous guidance and inspiration of Mr Harland Topham, formerly chief hydraulic design engineer at the English Electric Company Ltd.

REFERENCES

1. **Jaeger, C.** "Fluid transients in hydro-electric engineering practice". *Blackie, London* 1977

2. **Wylie, EB.** "Resonance in pressurised piping systems". *Journal of Basic Engineering, Trans. ASME,* December 1965

3. **British Standards Institution.** "BS 5500 Unfired fusion-welded pressure vessels" 1990.

4. **Grein, H.** "Vibration phenomena in francis turbines: their causes and prevention". *IAHR Symposium, Tokyo* 1980.

5. **Wilson, JD. & Tooth, AS.** "The support of unstiffened cylindrical pressure vessels". *2nd International Conference on Pressure Vessel Technology, ASME.* 1973

6. **Priddle, EK. & Durrans, RF.** "R2 assessment procedure (response and integrity of structures under vibrational loading) Volume 5: A procedure for assessing the fatigue life of non-defective structures". *Nuclear Electric internal report, TD/SID/REP/0069,* November 1992

Hydro turbine upgrades – new technology increases productivity

WILSON
Weir Engineering Services, Glasgow, UK
RUSSELL MASCE
American Hydro Corporation, Pennsylvania, USA

SYNOPSIS

This paper highlights the technology required to achieve successful upgrades of hydraulic turbines. The mechanical and hydraulic design processes utilised for the development of new turbine runners of maximum capacity and efficiency are detailed. Runner fabrication techniques developed to accurately produce high performance runners in a nominal delivery time are outlined. Three case studies which review typical successful turbine upgrades are presented.

INTRODUCTION

Hydroelectric power generation has long been a major contributor to power systems throughout the world. Installed over the past 100 years, much of the original equipment continues to operate reliably, although turbine performance is far below the performance of modern designs. In general, these existing units have been installed in the most attractive civil and hydrological settings even though the full hydraulic capacity of each site may not yet be utilised.

Increasing demand for clean, inexpensive electric energy challenges electric utilities to increase capacity. However, the cost to develop new sites has escalated dramatically. As an alternative to new installations, a much better economic return can be achieved by increasing the efficiency and capacity of existing power stations using modern technology to upgrade the hydroturbine equipment.

The most effective approach to hydroturbine upgrades evolved from the development of three-dimensional finite element computer codes, capable of analysing the flows and structures of existing components and developing modern designs for the critical components, primarily the runners. These codes have been in use over the past 20 years and have proven that older units can be made to produce from 10% to over 50% more capacity from the existing hydroturbines. Efficiency improvements have exceeded 10% and cavitation behaviour has also been improved. Weir Engineering Services associates, the American Hydro Corporation has been successful with over 250 such runner upgrades. This paper will discuss the initial phases of the upgrade process, the hydraulic and mechanical analysis and design. In addition to a presentation of the computer programs and engineering approach, three projects that benefited from these programs are also discussed. These projects proved to be successful not just because of the use of computer models, but also due to new processes enabling economic production of fabricated runners which are installed with a minimum of down time and tested to assure the value of the upgraded units. A brief discussion of the fabrication process will also be included in the presentation.

HYDRAULIC ANALYSIS

The initial step in a successful upgrade project is to conduct a hydraulic study that considers the details of all of the existing water passageways as well as the details of the future water supply conditions of the hydroelectric facility. The accuracy of the upgrade study depends on the accuracy of information provided in this initial study. If detailed drawings are not available, it is necessary to measure the shapes and dimensions of the spiral case, stay vanes, wicket gates, runner, seals and draft tube. The available water flow, including its relation to headwater and tailwater elevations, is used to determine potential capacity, while efficiency improvements and cavitation behaviour are determined by detailed hydraulic analysis.

A primary factor in developing new runner designs for existing hydro turbines is accurately predicting the distribution of water velocity and associated blade pressures such that the runner has optimum efficiency and cavitation performance. The "AHRDS" computer software system accomplishes two major functions. Developed by American Hydro engineers, "AHRDS" provides a computer interactive design environment with complete flexibility for runner shape definition. The geometry generation portion provides rapid runner hardware design. Figure 1 presents a typical runner geometry developed using "AHRDS". The fluid flow analysis tests the hydraulic performance of the runner geometry to determine power, efficiency and cavitation resistance. Through a series of design and analysis steps, a runner design can be optimised to provide the best possible hydraulic performance.

The fluid flow analysis provides a fully three-dimensional finite element calculation of water velocity, pressure, and energy for the wicket gates or for the runner. Figure 2 shows the finite element grid for a typical Francis runner. Due to the runner symmetry only one blade passage is modelled. Figures 3 and 4 present the predicted velocity and pressure fields along the suction side of the blade. Subsequent boundary layer calculations, using the momentum integral method with a finite difference technique, provide an estimate of the skin friction and thus, the efficiency.

Over the past 15 years this style of analysis has proven to be very accurate in determining turbine power and cavitation performance as shown by extensive model testing and field performance. In essence, the computer becomes the model test stand, thereby, fully testing a runner design before it is put into service. Through the computer testing of custom designed runners expensive physical model tests can often be eliminated. The versatility of "AHRDS" allows one to analyse individual flow regimes of all design shapes and sizes. This is particularly important for the optimisation of replacement runner designs. This system is unencumbered with historical design concepts imposed by original equipment designers and allows complete freedom in developing upgraded runners for each application. The result of this approach is becoming more evident in the increasingly active rehabilitation market where most utilities have elected to specify computer design analysis for their upgrade projects.

STRUCTURAL ANALYSIS

In addition to the hydraulic improvements that result from an upgrade, sometimes runners need replacement due to structural failures. When cracks or cavitation damage require protracted maintenance expenses, a runner upgrade provides the compound benefits of maintenance and outage expense reduction and increased productivity beyond that which exists. An upgraded runner must be analysed to assure that it will operate safely and that the unit structure will provide a reliable extended service life.

A general purpose structural analysis system, "AHSTRUCT", is capable of handling static and dynamic analysis of beams and shafting systems (one dimensional), plates, shells and general axisymmetric structures (two dimensional) and general three dimensional structural configurations, such as solid castings, trusses, reinforced concrete, weldments, etc. The two and three dimensional models are formulated using cubic, sub-parametric elements including all displacements and their first derivatives which represent the slopes and strains. Therefore, the matrix solution of the finite element mathematical model directly supplies the stress and strain elasticity relations giving very accurate assessment of even local surface stress concentrations with minimum structure discretization. This is a prerequisite to any engineering design whose cyclic loading and related fatigue life determines strength requirements.

The analytical values of local cyclic strains can then be post-processed using local cyclic strain based fatigue analysis and fracture mechanics software to determine life, inspection criteria and safety margin. The finite element analysis consider point and pressure loads as well as centrifugal and gravitational body generated forces and thermal loadings. The structure's stability is modelled by imposing constraints or ground stiffness to any of the deformation degrees of freedom. Tied degrees of freedom allow assessment of intercomponent loading and behaviour such as in press fits, keys and keyways, flange joints, etc.

Pre-processors help to generate and verify the mathematical model using graphics and printed diagnostics. Post-processors are used to combine linearly the separate load cases and display the stress, deformation, reactions or mode shape results graphically or in printed form. Accurate yet inexpensive, non-linear analysis can be made of structures with variable boundary conditions using a special post-processing technique.

RUNNER MANUFACTURING

The technology developed by American Hydro to manufacture turbine runners relies extensively on fabrication techniques involving the use of high quality stainless steel plate. Traditionally, the approach has been to supply a one-piece cast runner, or a runner fabricated from cast components. The one-piece cast approach precludes the ability to provide a runner with blade shapes that are highly accurate. Furthermore, the use of castings in general presents several disadvantages which are overcome with the plate fabrication approach. The following should be considered:

1. The quality associated with stainless steel plate is inherently high. To approach the quality level of plate, an extensive amount of non-destructive examination and upgrading must first be carried out on castings.
2. Any volumetric inspection performed on plate, with its uniform thickness, is much more reliable, as compared to castings with sections of varying thickness.
3. Experience has shown that manufacturing a runner from plate can generally be accomplished in a shorter period of time than when castings are involved. The risk of late delivery is reduced since a greater percentage of the work is performed under the control of the manufacturer. Casting defects, which can open up during finish machining operations, are also eliminated.
4. Plate is readily available from several domestic sources, while the number of casting suppliers is dwindling.

Most of the runners supplied by American Hydro to date have been fabricated with plate material conforming to ASTM A240 UNS S30403. This is an excellent material for most fresh water applications, providing good corrosion and cavitation resistance. For non stress or shape critical components (such as the runner crown) the cast equivalent, ASTM A744 Grade CF3, is sometimes used.

American Hydro utilises in house developed software to automate the manufacturing process. The runner blade, crown and band shapes are segmented and developed into flat pieces which are automatically nested onto a plate to minimise scrap loss. The CNC plasma cutting program is automatically created and sent via a DNC link to the plasma cutting machine which cuts these segments. Fabricated, CNC-machined forming dies are produced on a parallel path.

After hot-forming, the butt joint weld geometry is applied by a machining operation to the plate segments, which are subsequently assembled and loaded onto a welding positioner. Positioning the piece for flat welding optimises the welding deposition rate and quality. Premachining of the waterpassage surfaces of the crowns and bands are carried out after welding.

The flat blade shapes are machined to obtain the design airfoil thickness distribution. Hot forming of the blades between precision machined dies is followed by machining of the weld geometry at the intersections with the band and crown. Blades manufactured in this manner have been found to be superior to those machined from castings from both aspects of dimensional accuracy, consistency from blade to blade and structural integrity.

The blades are positioned between the crown and band using fixtures which accurately establish their locations. Flux-cored arc, shielded-metal arc, and gas-tungsten arc welding processes are utilised to produce the full penetration welds used in the high stress areas joining the buckets to the band and crown. Following both dye penetrant and ultrasonic examinations of the weld joints the runner is finish machined. The final step is to balance the runner on a high speed two plane precision dynamic balancing machine.

CASE STUDIES

Yale Hydroelectric Project

The Yale Hydroelectric Project, located on the Upper Lewis River near Mount St. Helens in Washington State, is an important part of the Lewis River Hydroelectric System. This power plant houses two generating units, each with an original rating of 54 MW, that have been in continuous operation since 1953. Supply water is retained by an earthfill dam that forms the scenic upper reservoir with 402,000 acre feet of total storage capacity.

Taking advantage of modern technology, the utility has undertaken the task of replacing the original turbine runners with the objective to optimise annual energy generation. The initial turbine upgrade analysis using the "AHRDS" program demonstrated that significant gains could be made. Confidence in these techniques led the customer to procure the modern design runners with no physical model test. The new runners were designed and manufactured by American Hydro and installed in 1996.

To maximise capacity and efficiency, the turbine discharge ring was modified to accept a deeper runner. Figure 5 shows a cross section of the turbine unit with the upgraded runner.

The mechanical design of the upgraded replacement turbine runner for Yale was verified using advanced structural analyses which accurately identify deflections, maximum stresses, mode and frequencies of vibration, fatigue life and material flaw acceptance criteria. These analyses using "AHSTRUCT" were also used to specify appropriate connection weld geometries. Modal studies were made to identify the operational mode shapes and respective natural frequencies to make sure that resonant vibration would not be excited by known forcing frequencies.

Figures 6 and 7 show the deflections of the 17 bladed runner design magnified at 60 X in dashed line relative to the unloaded runner in solid line for maximum power generation and runaway conditions respectively. The water pressure distributions were obtained from the results of the three-dimensional finite element analysis of the fluid flow through these runners, Figure 4. The mathematical model uses cubic subparametric solid finite elements which accurately evaluate the stress risers directly on the blade fillet toe. Such accurate assessments of local cyclic stress is prerequisite to meaningful fatigue life and crack growth evaluation.

The stress data along with tested material properties of the stainless steel blade material provide the necessary input to American Hydro's proprietary and proven fatigue life and fracture mechanics analytical software. The results of the fatigue studies predict a long, reliable life for the new runner. Both the fatigue and fracture mechanics analyses acknowledged the presence of residual weld stresses equal to 90% of the 304 stainless steel's yield stress to ensure that the runners will operate safely for their extended life.

Field performance testing, using a multi-path sonic method to measure flow, demonstrated a substantial increase in unit characteristics. The output capacity was increased by 30% and the efficiency was increased by nearly 10%. Figure 8 presents these field results. In addition to smooth running operation, the units have demonstrated excellent cavitation behaviour.

Stonebyres

Stonebyres Power Station is the smaller of two stations in the Falls of Clyde Scheme owned and operated by Scottish Power plc.

The stations contain two Francis units rated at 3 MW at 30 metre head. The units were commissioned in 1926/27 and have been in commercial operation some 70 years.

In 1995 Scottish Power placed an order with Weir Engineering Services for a replacement runner for the Number 2 machine. An inclusive part of the order was Pre and Post replacement efficiency testing and the site work associated with runner removal and refitting.

By using "AHRDS" it was determined that a new runner design could be manufactured that would increase the unit output while improving cavitation behaviour.

The Stonebyres runner was fabricated completely from stainless steel material. ASTM A240 type 304L was chosen for the buckets for its advantages with respect to dimensional control, corrosion resistance, field repair welding and material toughness. To facilitate an accelerated delivery requirement, the crown and band were spun cast from ASTM A744 Grade CF3, which has equivalent properties to 304L. The material selections were made based on the mechanical analysis, using "AHSTRUCT", as well as considerable experience with runner fabrication. The buckets were fully machined and then hot pressed into their final form, fully meeting the physical properties of the ASTM specification and virtually free from residual stresses. The runner design for Stonebyres was made with a maximum stress level at less than one third of the material's yield strength.

A final consideration for the use of 304L was that matching strength welds are obtained using austentic electrodes at ambient temperatures (approximately 60 degrees F). The best properties are developed in the as-welded condition because the material is not hardenable. Should any repair welding be required on the runner after it is installed, in-situ welding is easily accomplished.

The 60 inch turbine runner was designed, fabricated and delivered in only four months after contract award. Figure 9 shows the field test results. An average relative efficiency increase of 10% was recorded while the relative efficiency at full load was increased by 20%.

Bull Run

The Bull Run Power Plant near Sandy, Oregon houses horizontal shaft hydro units rated at 5 MW at 320 feet head. Originally installed in 1912, the Francis turbines were operating at reduced efficiency levels due to an antiquated hydraulic design of runners and wicket gates. The utility decided on a turbine upgrade to increase energy generation through more efficient use of the available water. While the cost to absolute efficiency test these units was prohibitive, American Hydro guaranteed the upgraded unit output and a weighted average efficiency increase. These guarantees could then be verified by relatively inexpensive index tests. American Hydro designed the runners to maximise the discharge area and minimise the expansion loss into the draft tube. The new custom designed, fabricated wicket gates and runner provided a relative efficiency improvement of nearly 10% and a 7% increase in maximum capacity. The results of the index test are shown in Figure 10. The payback for the investment was realised in less than three years.

CONCLUSION

Perhaps the most economical forms of new electric power come from the upgrade of existing hydroelectric plants. Frequently the best hydraulic sites have been in operation the longest and therefore utilise the oldest of designs. Without the financial burden of constructing new civil structures, upgrade projects have proven to be economical, quickly implemented, environmentally friendly contributors to utility system demands for added capacity. The incremental increases can range from a nominal 10% to over 50% of the existing capacity.

With an accurate hydraulic assessment, the benefits of upgrading old units with modern runners and perhaps modifications to other components can be determined with assurance. Extensive model and field testing and over 250 successful runner upgrades in the last ten years has confirmed the accuracy of the analytical tools discussed here, "AHRDS" and "AHSTRUCT". These tools help ensure that the financial assessment will provide reliable direction for funding. Once the project is considered feasible, the process continues with detailed hydraulic and mechanical analysis to develop the final design for optimised output, structural integrity and renewed life extension for the complete unit. Manufacturing to strict tolerance standards and quality are required to achieve the predicted performance. A significant economic benefit is achieved using fabrication techniques which reduce manufacturing times and co-ordination with site work to minimise downtime for the installation of the new runner and components.

Utility companies continue to recognise and implement the significant benefits gained by applying new design techniques developed specifically for efficiency and capacity increases to existing hydroelectric units. Modern computerised designs are instrumental in optimising the financial benefits.

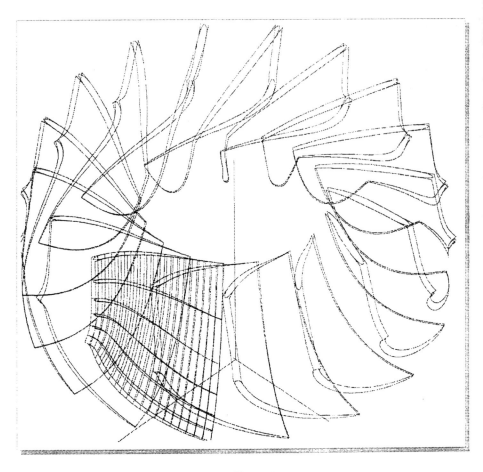

Figure 1
Typical Three-dimensional View of "AHRDS" Runner

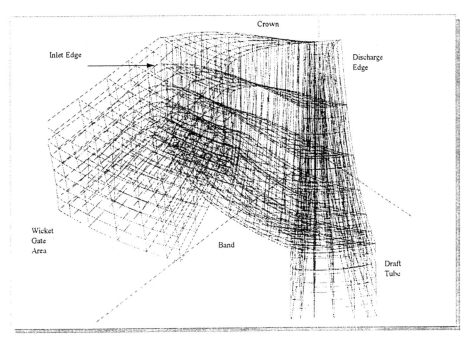

Figure 2
Typical "AHRDS" Finite Element Grid

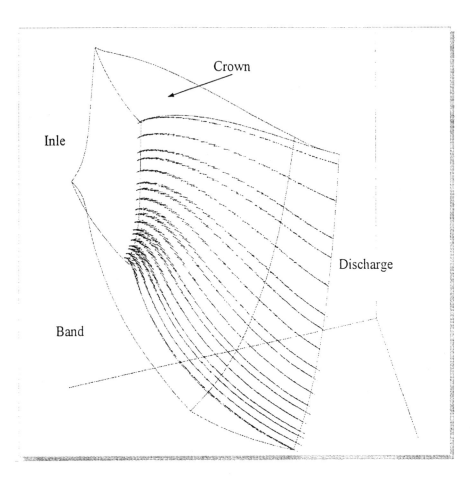

Figure 3
Yale Upgraded Runner
"AHRDS" Velocity Field - Low Pressure Surface

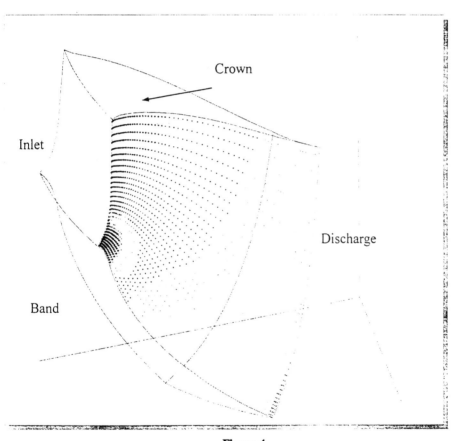

Crown

Inlet

Discharge

Band

Figure 4
Yale Upgraded Runner
"AHRDS" Pressure Distribution - Low Pressure Surface

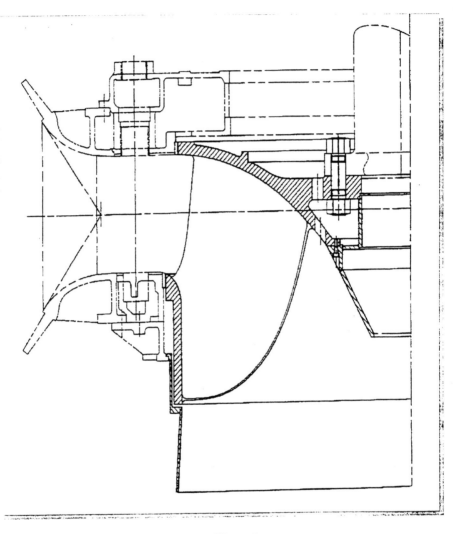

Figure 5
Yale Hydroturbine Cross Section

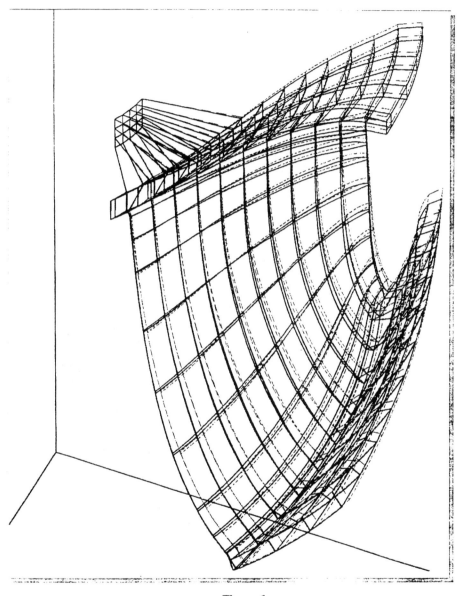

Figure 6
Yale Upgraded Runner - Maximum Power
Scale = .302   Magnification x 60

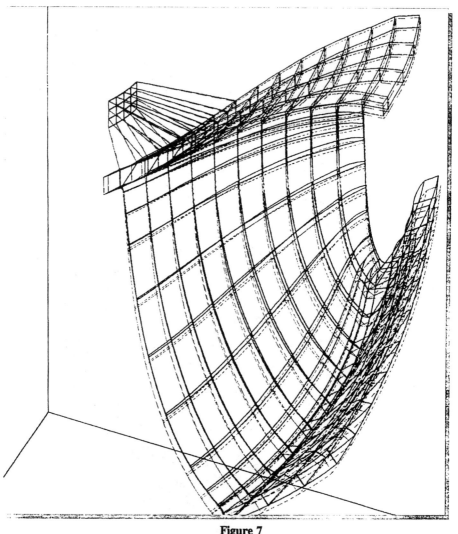

Figure 7
Yale Upgraded Runner - Runaway
Scale = .296   Magnification x 60

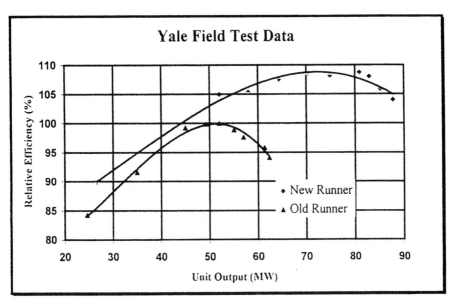

Figure 8

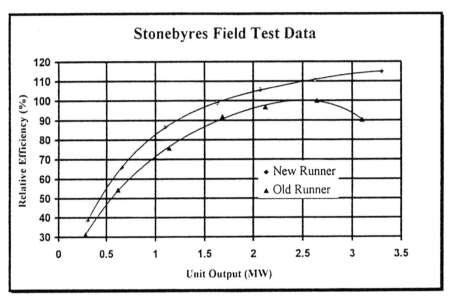

Figure 9

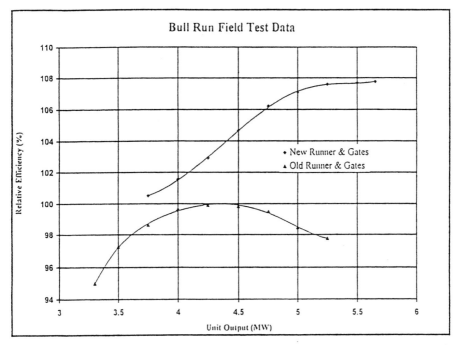

Figure 10

Authors' Index